MICROELECTRONICS AND OPTOELECTRONICS

The 25th Annual Symposium of Connecticut Microelectronics and Optoelectronics Consortium (CMOC 2016)

SELECTED TOPICS IN ELECTRONICS AND SYSTEMS

Editor-in-Chief: **M. S. Shur**

Published*

Vol. 49: Spectral Sensing Research for Surface and Air Monitoring in Chemical, Biological and Radiological Defense and Security Applications
eds. Jean-Marc Theriault and James O. Jensen

Vol. 50: Frontiers in Electronics
eds. Sorin Cristoloveanu and Michael S. Shur

Vol. 51: Advanced High Speed Devices
eds. Michael S. Shur and Paul Maki

Vol. 52: Frontiers in Electronics
Proceedings of the Workshop on Frontiers in Electronics 2009 (WOFE-2009)
eds. Sorin Cristoloveanu and Michael S. Shur

Vol. 53: Frontiers in Electronics
Selected Papers from the Workshop on Frontiers in Electronics 2011 (WOFE-2011)
eds. Sorin Cristoloveanu and Michael S. Shur

Vol. 54: Frontiers in Electronics
Advanced Modeling of Nanoscale Electron Devices
eds. Benjamin Iñiguez and Tor A. Fjeldly

Vol. 55: Frontiers in Electronics
Selected Papers from the Workshop on Frontiers in Electronics 2013 (WOFE-2013)
eds. Sorin Cristoloveanu and Michael S. Shur

Vol. 56: Fundamental & Applied Problems of Terahertz Devices and Technologies
Selected Papers from the Russia–Japan–USA Symposium (RJUS TeraTech-2014)
ed. Michael S. Shur

Vol. 57: Frontiers in Electronics
Selected Papers from the Workshop on Frontiers in Electronics 2015 (WOFE-15)
eds. Sorin Cristoloveanu and Michael S. Shur

Vol. 58: Fundamental and Applied Problems of Terahertz Devices and Technologies
Selected Papers from the Russia-Japan-USA-Europe Symposium (RJUSE-TeraTech 2016)
by Maxim Ryzhii, Akira Satou and Taiichi Otsuji

Vol. 59: Scaling and Integration of High Speed Electronics and Optomechanical Systems
eds. Magnus Willander and Håkan Pettersson

Vol. 60: Microelectronics and Optoelectronics
The 25th Annual Symposium of Connecticut Microelectronics and Optoelectronics Consortium (CMOC 2016)
eds. F. Jain, C. Broadbridge and H. Tang

*The complete list of the published volumes in the series can be found at
http://www.worldscientific.com/series/stes

Selected Topics in Electronics and Systems – Vol. 60

MICROELECTRONICS AND OPTOELECTRONICS

The 25th Annual Symposium of Connecticut Microelectronics and Optoelectronics Consortium (CMOC 2016)

University of Connecticut, USA   6 April 2016

Editors

F. Jain
University of Connecticut, USA

C. Broadbridge
Southern Connecticut State University, USA

H. Tang
Yale University, USA

NEW JERSEY • LONDON • SINGAPORE • BEIJING • SHANGHAI • HONG KONG • TAIPEI • CHENNAI

Published by

World Scientific Publishing Co. Pte. Ltd.
5 Toh Tuck Link, Singapore 596224
USA office: 27 Warren Street, Suite 401-402, Hackensack, NJ 07601
UK office: 57 Shelton Street, Covent Garden, London WC2H 9HE

British Library Cataloguing-in-Publication Data
A catalogue record for this book is available from the British Library.

Selected Topics in Electronics and Systems — Vol. 60
MICROELECTRONICS AND OPTOELECTRONICS
The 25th Annual Symposium of Connecticut Microelectronics and Optoelectronics Consortium (CMOC 2016)

Copyright © 2017 by World Scientific Publishing Co. Pte. Ltd.

All rights reserved. This book, or parts thereof, may not be reproduced in any form or by any means, electronic or mechanical, including photocopying, recording or any information storage and retrieval system now known or to be invented, without written permission from the publisher.

For photocopying of material in this volume, please pay a copying fee through the Copyright Clearance Center, Inc., 222 Rosewood Drive, Danvers, MA 01923, USA. In this case permission to photocopy is not required from the publisher.

ISBN 978-981-3232-33-4

Printed in Singapore

Preface

The first symposium was held in 1992 at the United Technologies Research Center, East Hartford, CT. Thereafter, year-after-year, a team of experts working on the cutting edge of Microelectronics and Optoelectronics would get together to present and share their research results on the subject. The 25th annual symposiums Connecticut Microelectronics and Optoelectronics Consortium (CMOC), organized by a team of seven academic institutions and about eighteen companies, was held on April 6, 2016 at the University of Connecticut. In this symposium, we attracted authors from all over the United States. The key note speakers included representative from IMEC, Belgium.

The focus of the Symposium, as its name suggests, is primarily micro/nano-electronics and optoelectronics/nano-photonics. In this volume, however, we have contributed papers ranging from biosensors/nano-biosystems, emerging technologies, and applications.

Enabling materials research involving growth and characterization of novel devices such as multi-bit nonvolatile random access memory with fast erase, high performance circuits, and their applications in developing new high-speed systems is the focus of this special issue.

Some papers in this volume focus on emerging nanoelectronic devices including topological insulators, spatial wavefunction switching (SWS) FETs as compact high-speed 2-bit SRAM circuits, quantum dot channel (QDC) FETs.

Pixel characterization of protein-based retinal implant is the topic of one paper. While another paper demonstrates using carbon nanofibers/nanotubes for electrochemical sensing. Carbon nanotube synthesis from block copolymer deposited catalyst is the theme of another paper. A low-power and low-data-rate (100 kbps) fully integrated CMOS impulse radio ultra-wideband (IR-UWB) transmitter for biomedical application is presented in a complementary paper.

Fundamental work on critical layer thickness in ZnSe/GaAs and other material systems impacts electronic and photonic devise integrating mismatched layers. Another paper investigates linearly graded GaAsP-GaAs system with emphasis on strain relaxation. Still another paper analyzes multiple junction solar cells using semiconductors with different energy gaps.

In the area of cyber security, two papers present encrypted electron beam lithography fabricated nanostructures for authentication and nano-signatures for the identification of authentic electronic components.

In summary, papers presented in this volume involve various aspects of high performance materials and devices for implementing high speed electronic systems.

Guest Editors:

F. Jain (University of Connecticut)
C. Broadbridge (Southern Connecticut State University)
H. Tang (Yale University)

Contents

Preface   v

Fabrication of Robust Nano-Signatures for Identification of Authentic Electronic Components and Counterfeit Avoidance   1
    K. Ahi, A. Rivera, A. Mazadi and M. Anwar

Progression of Strain Relaxation in Linearly-Graded $GaAs_{1-y}P_y/GaAs$ (001) Epitaxial Layers Approximated by a Finite Number of Sublayers   11
    T. Kujofsa and J. E. Ayers

Carbon Nanotubes, Nanofibers and Nanospikes for Electrochemical Sensing: A Review   25
    A. S. Shanta, K. A. Al Mamun, S. K. Islam, N. McFarlane and D. K. Hensley

Spatial Wavefunction Switched (SWS) FET SRAM Circuits and Simulation   37
    B. Saman, P. Gogna, E-S. Hasaneen, J. Chandy, E. Heller and F. C. Jain

Carbon Nanotube Synthesis from Block Copolymer Deposited Catalyst   49
    K. Woods, J. Silliman and T. C. Schwendemann

Dynamical X-Ray Diffraction Analysis of Triple-Junction Solar Cells on Germanium (001) Substrates   55
    F. A. Althowibi and J. E. Ayers

Pixel Characterization of a Protein-Based Retinal Implant Using a Microfabricated Sensor Array   69
    J. A. Greco, L. A. L. Fernandes, N. L. Wagner, M. Azadmehr, P. Häfliger, E. A. Johannessen and R. R. Birge

A Low-Power Low-Data Rate Impulse Radio Ultra-Wideband (IR-UWB) Transmitter   89
    I. Mahbub, S. Shamsir and S. K. Islam

Multi-Bit NVRAMs Using Quantum Dot Gate Access Channel   99
    M. Lingalugari, P.-Y. Chan, J. Chandy, E. Heller and F. Jain

Quantum Dot Channel (QDC) Field Effect Transistors (FETs) Configured as Floating Gate Nonvolatile Memories (NVMs)   113
    J. Kondo, M. Lingalugari, P. Mirdha, P.-Y. Chan, E. Heller and F. Jain

Denoising and Beat Detection of ECG Signal by Using FPGA   125
    D. Alhelal and M. Faezipour

Encrypted Electron Beam Lithography Nano-Signatures for Authentication   137
    K. Ahi, A. Rivera and M. Anwar

Topological Insulators: Electronic Structure, Material Systems and Applications   151
    P. Sengupta

Mosaic Crystal Model for Dynamical X-Ray Diffraction from Step-Graded $In_xGa_{1-x}As$ and $In_xAl_{1-x}As/GaAs$ (001) Metamorphic Buffers and Device Structures   175
    P. B. Rago and J. E. Ayers

Critical Layer Thickness: Theory and Experiment in the ZnSe/GaAs (001) Material System   187
    T. Kujofsa and J. E. Ayers

Author Index   197

Fabrication of Robust Nano-Signatures for Identification of Authentic Electronic Components and Counterfeit Avoidance

Kiarash Ahi[*], Abdiel Rivera, Anas Mazadi and Mehdi Anwar

Department of Electrical Engineering,
University of Connecticut,
Storrs, CT 06268, USA
[*]kiarash.ahi@uconn.edu

In this paper, a novel approach for marking integrated circuit packages with authentication nano-signatures is introduced. In this work, the signatures patterns are fabricated using electron beam lithography. Moreover, the robustness of these signatures against aging and humidity is investigated. A recipe comprising image processing techniques and measurement of similarity indices has been developed. These signatures are proposed to be fabricated at the manufacturer side of the supply chain. Then, they are decoded at the consumer end. Thus, robustness against ambient environment and aging is a requirement for these signatures to survive in the global supply chain. Calculated Mean Square Error and Structural SIMilarity Index confirmed that the reflected patterns of the signatures remain unchanged against aging and humidity.

Keywords: counterfeit avoidance; electron beam lithography; engineered nanostructures; image processing, SSIM.

1. Introduction

Since the early civilizations, counterfeiting of customers' goods has been always a great hindrance in front of innovation and protection of intellectual property. As the technology becomes more sophisticated, concerns about the early failure of the critical equipment grow. These critical systems consist of thousands of electronic and mechanical components. Consequently, having one counterfeit component results in the failure of the entire system. Such failures are responsible for numerous amount of deaths each year [1]. Overdesigning the systems by adding modular redundancies adds to the costs, volumes and weights of the systems, which are crucial factors for critical systems such as aerospace and transportation systems, and thus it is preferred to enhance the reliability of a system by using reliable components other than adding redundancies [2], [3].

Electronics components travel all around the world before they are assembled into different systems. Each component is made in different parts of the world. It is more economical for the companies to obtain their needed components from various foundries.

[*]Corresponding author.

However, tracking all the components in such global market is not possible. Thus, counterfeiters have more freedom to inject their products in the supply chain nowadays [4]–[6].

The financial loss of supply chains due to counterfeit electronic components is reported to be higher than $10 billion per year in 2010 [7]. This loss would be tremendous by considering the equipment failures because of malfunctions of counterfeit electronic components. In a more general view, the counterfeiting and pirated products have imposed catastrophic damages on the economies worldwide. The total value of counterfeit and pirated products in case of G20 countries are reported to be up to $650 billion in 2008 and it is predicted to rise to $1,770 billion in 2015 [8]. As the electronic industry is growing fast, counterfeit electronic components bring more revenue to the counterfeiters and thus, more electronic counterfeit components are injected into the market. As Fig. 1 indicates, the counterfeit electronic component incident reports have closely followed the global semiconductor revenue so far. Consequently, since a sharp rise in electronic components market is predicted, a high rise in counterfeit components is expected. As customers and suppliers are making more advanced methods for detecting counterfeit components, the counterfeiters make their techniques more sophisticated [9]. Consequently, innovative approaches for detecting counterfeit components are needed to be continually devised [10]–[15].

Most of the methods for detection of counterfeit ICs have been based on physical inspections and electric tests [10], [16]–[18]. As counterfeiters are making their approaches more advanced, the need for incorporating signatures on the electronic components is arising. Such signatures are not only needed to be invisible from counterfeiters but also not to be replicable by counterfeiters. Towards this aim, engineered nanostructures (ENS) have been implemented on the surfaces of integrated circuits (ICs) and approaches to detect them without the need for sophisticated costly machines are developed [19], [20].

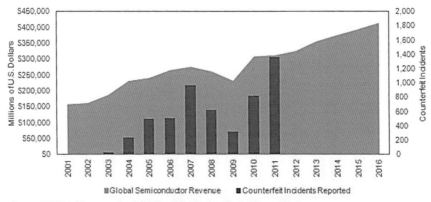

Fig. 1. Global semiconductor revenue correlated with reports of counterfeit parts [3].

This paper introduces a novel approach for implementation and detection of ENS for identification of authentic electronics components and any other products in general. It is believed that the proposed technology offers a tool (1) which allows identification of good ICs, already been captured and in post-design phase, and (2) which makes it impossible to resurface and re-introduced in the market since sanding of the surface destroys the ENS [21].

The application of these signatures implies that they need to be robust against aging and ambient environment effects. These signatures are required to survive the long journey in the supply chain from the manufacturers to the customers. In this work, a recipe comprising image processing and measuring similarity indices is developed to investigate the endurance of the fabricated nano-signatures as they are exposed to the high humidity and aging process. Calculated Mean Square Error (MSE) and Structural SIMilarity Index (SSIM) proved that the reflected patterns of the signatures remain unchanged against aging and humidity.

This paper is organized as follows: the second section introduces the procedures for the fabrication of ENS using electron beam lithography and the concepts of matrix patterns which can carry coded signatures. The third section provides a recipe for investigation of the robustness of ENS against effects of aging and the ambient environment. The provided results in this section prove that the reflected patterns of the signatures remain unchanged against aging and humidity.

2. Engineered Nanostructures

The incorporation of ENS may either be carried out by transferring the pattern to a thin metallic layer on the IC capping material or can be a direct write on nano-imprint. Engineered patterns ranging in size from nanometers to submicron dimensions are incorporated on the IC capping material as a tool for the identification of authentic ICs. Figure 2 shows a cell and a 3 × 3 array of an engineered nanostructure that should be recognized as a standard nano-signature structure imparting the properties of an encrypted signature matrix. As this figure also indicates, the structure is fabricated on the surface of a commercially available IC using Electron Beam Lithography (EBL) followed by Au sputtering.

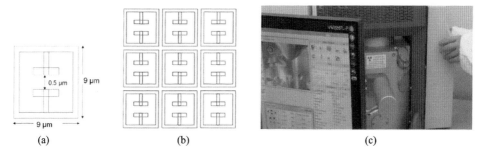

Fig. 2. (a) An ENS cell, (b) A 3 × 3 array of ENS, (c) The EBL system in UConn used for fabrication of the ENS.

Fig. 3. The experimental setup and the presence of a second reflection introduced by the encrypted signature matrix.

The ENS is invisible to the naked eyes but is detectable by using inexpensive and fast optical probing such as shining a laser pointer on the surface of the IC as shown in Fig. 3. The figure also shows the experimental setup that includes a laser pointer, a focusing lens, the device under test and the reflections.

3. Robustness of the ENS against Aging and Ambient Effects

An electronic component travels all around the world in the supply chain until being assembled in a system. The component is exposed to the different ambient environment until it is received by the customer. This journey also takes time and the ENS is aged. Consequently, the application of the proposed technology requires the structures to remain robust against aging. Moreover, the second optical reflection that is correlated to the signature matrix must remain invariant during all the changes in ambient over time. Figure 4 depicts the journey of the ENS in the supply chain, between the manufacturer and the customer.

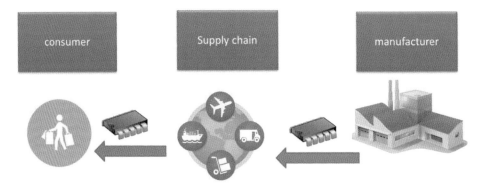

Fig. 4. The journey of a component in the supply chain.

To investigate the robustness of the fabricated structures, the ENS is exposed to different ambient environment and aging. After these exposures, different similarity indices are computed. Figure 5(a) depicts the ENS reflection under different ambient effects. For the case of determination of the effects of the ambient environment on the ENS, the images of the laser reflections from the ENS (1) after the ENS has been exposed to high moisture and (2) once the ENS has been stored for three months in the room environment, are obtained. The obtained images are then compared to the image of the laser reflection from the fresh ENS by SSIM, MSE and Euclidean distance (ED). Also, an image processing recipe is developed to make the images prepared for applying the mentioned similarity metrics.

As the first step of this recipe, since the colors of the obtained images are not playing a role, the colors are filtered. Then, in the second step, before analyzing the images with SSIM, MSE, and ED, since the reflection from the ENS is brighter than the background, for suppressing the background noise, pixels with the luminance of lower than 0.2 are filtered. The resulting images and corresponding histograms after these two filtrations are illustrated in Fig. 5. Then the overall similarities of the images in Fig. 5(d) are computed by using different metrics, namely SSIM, ED, and MSE. The reason that SSIM is chosen as a measure in this work is the fact that this metric has been developed as an index to determine the quality of a distorted image compared to its original counterpart [22], [23]. Here the quality of the exposed ENS to the ambient effects and aging is to be compared to the fresh ENS and thus SSIM can provide a good measure for this purpose. The algorithm of calculation of this metric is developed based on the principles of human visual systems. The human visual system is mainly focused on detection of the structures. Two other parameters which are effective on the human visual perception of image fidelity and quality are luminance and contrast. Equation (1) indicates that SSIM index takes the three mentioned parameters into account, namely luminance, contrast, and structure.

$$SSIM(x,y) = [l(x,y)]^{\alpha} \cdot [c(x,y)]^{\beta} \cdot [s(x,y)]^{\gamma} \tag{1}$$

Where l, c and s are representing similarities in terms of luminance, contrast, and structure between two images respectively. $\alpha > 0$, $\beta > 0$ and $\gamma > 0$ determines the relative importance of these three components. Values between 0 and 1 are assigned to l, c and s, the higher the value, the higher the similarity of the corresponding parameter between the two compared images. The output is a value between 0 to 1. The higher the value is, the more similar the images are. For this work, only the structural shape of the ENS determines the authentication of the product. Consequently, structural similarity is to be measured. For this purpose, the effect of the contrast and luminance are to be eliminated from Eq. (1). By substituting $\alpha = \beta = 0$ the structural similarity from the SSIM index can be extracted as:

$$SSIM(x,y) = s(x,y) = \frac{\sigma_{xy} + C}{\sigma_x \sigma_y + C} \tag{2}$$

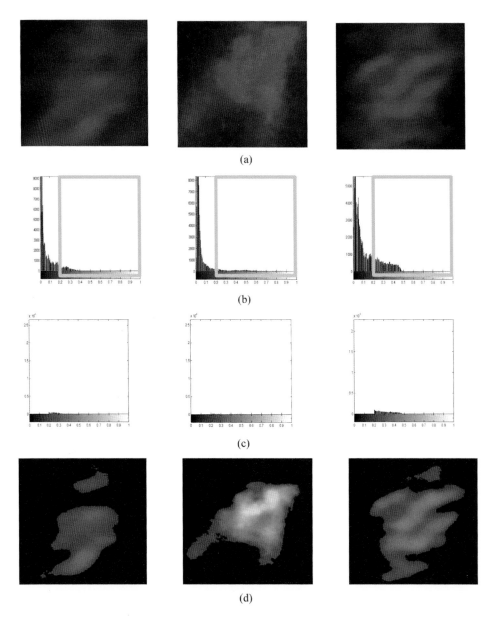

Fig. 5. (a) Original images of the reflections of ENS (from left to right): without being exposed to ambient environment effects, after three months and after being exposed to high moisture. (b) Histogram of the original images in (a); the yellow windows show the filter window. (c) Histograms of the images after being filtered. (d) Images after filtering and removing colors.

Table 1. Results of similarity measurements with SSIM, MSE and ED.

Row #	The two images which are being compared to each other and the number of figure in with their corresponding fft2 is depicted		SSIM	MSE	ED
	First image	Second image			
1	High moisture	Original	0.9988	0.0077	18.3457
2	After a few days	High moisture	0.9969	0.0198	29.4204
3	After a few days	Original	0.9963	0.0237	32.1955
4	After a few days	3 months old	0.9802	0.1264	74.4005
5	3 months old	Original	0.9791	0.1333	76.4078
6	High moisture	3 months old	0.9779	0.1408	78.5268

(Similarity: The highest → The lowest)

In a more general sense, SSIM index is a promising measure to determine the structural similarity. The results of measurement of the degradation of ENS by the above index are shown in Table 1. The high values of SSIM confirm the robustness of the ENS. A logic between the results in the Table is observed. As the ENS is aged its similarity to the fresh ENS decreases: the few days aged ENS is more similar to the fresh ENS than the 3-month aged ENS is. Also, the two aged ones are more similar to each other than to the fresh one. As expected, the less similarity would be between the exposed one to the high humidity and the 3-month aged one. It is also seen that the humidity has less degradation effect than aging. It is desirable to observe degradation in the ENS as a measure of aging. The object is exposed to the high humidity in the supply chain and it must remain unchanged. However, for detecting the recycled aged components, observing the degradation in the ENS is an advantage.

For confirming the result with a completely different metric, MSE which has been one of the most well-known metrics to measure differences between two signals is also used [24]. The MSE between two arbitrary finite length discrete signals (which can represent two visual images) $x = \{x_i \mid i = 1, 2, \cdots, N\}$ and $y = \{y_i \mid i = 1, 2, \cdots, N\}$ is defined by Eq. (3).

$$\text{MSE}(x,y) = \frac{1}{N} \sum_{i=1}^{N} (x_i - y_i)^2 \qquad (3)$$

Euclidean distance (ED) can be considered as the ordinary distance between two points and it is given by the Pythagorean formula. Consequently, for the two arbitrary finite length discrete signals x and y, ED can be calculated by Eq. (4).

$$\text{ED}(x,y) = \sqrt{\sum_{i=1}^{N} (x_i - y_i)^2} \qquad (4)$$

The results of the degradation measurements using MSE and ED are also given by Table 1. Interestingly, the results from the SSIM are the same as the results from the MSE and ED, although the SSIM is based on a completely different logic to determine the similarities. This consistency between SSIM and MSE confirms the accuracy of the proposed approach. It is observed that calculated SSIM indices are very high, and MSEs are very small. It translates to the fact that the ENS are robust against aging and ambient effects.

4. Conclusion

In this paper, a method for fabrication of engineered nano-signatures was developed. Concepts of encoded matrices which can carry coded signatures were adopted to realize these signatures. The application of the proposed technology requires the structures to remain robust over time and under different ambient conditions. The fabricated structures were exposed to the different ambient environment and aging processes. Similarity indices were computed and it was observed that the ENS has not been degraded. In this way, the robustness of the engineered nanostructures against effects of ambient environment is proved.

Acknowledgment

The authors would like to acknowledge Center for Hardware Assurance and Engineering (CHASE) in UConn for supporting this work.

References

[1] Http://www.oecd.org/dataoecd/13/12/38707619.pdf, "OECD, The Economic Impact of Counterfeiting and Piracy."
[2] K. Ahi, "Control and Design of a Modular Converter System for Vehicle Applications," Leibniz Universität Hannover, Hannover- Germany, 2012.
[3] K. Ahi, "Modeling and control of a multiphase modular high frequency converter/ inverter for vehicle applications," *Math. Probl. Eng.*, 2017.
[4] F. A. Tillman, C. L. Hwang and W. Kuo, "Determining Component Reliability and Redundancy for Optimum System Reliability," *IEEE Trans. Reliab.*, vol. R-26, no. 3, pp. 162–165, 1977.
[5] U. S. Congress, "Ike Skelton National Defense Authorization Act for Fiscal Year 2011." [Online]. Available: http://www.gpo.gov/fdsys/pkg/BILLS-111hr6523enr/pdf/BILLS-111hr6523enr.pdf.
[6] "Defense Industrial Base Assessment: Counterfeit Electronics," 2010.
[7] J. M. Radman and D. D. Phillips, "Novel Approaches for the Detection of Counterfeit Electronic Components," *IN Compliance Magazine*, no. October, 2010.
[8] Frontier Economics Ltd., "Estimating the global economic and social impact of counterfeiting and piracy," 2011.
[9] IHS, "Counterfeit-Part Risk Expected to Rise as Semiconductor Market Shifts into Higher Gear," EL SEGUNDO, Calif., 2012.
[10] K. Ahi and M. Anwar, "A Novel Approach for Enhancement of the Resolution of Terahertz Measurements for Quality Control and Counterfeit Detection," in *Diminishing Manufacturing Sources and Material Shortages (DMSMS)*, 2015.
[11] K. Ahi and M. Anwar, "Modeling of terahertz images based on x-ray images: a novel approach for verification of terahertz images and identification of objects with fine details beyond terahertz resolution," in *Proc. SPIE 9856, Terahertz Physics, Devices, and Systems X: Advanced Applications in Industry and Defense, 985610*, 2016, p. 985610.

[12] K. Ahi and M. Anwar, "Advanced terahertz techniques for quality control and counterfeit detection," in *Proc. SPIE 9856, Terahertz Physics, Devices, and Systems X: Advanced Applications in Industry and Defense, 98560G*, 2016, p. 98560G.
[13] K. Ahi and M. Anwar, "Developing terahertz imaging equation and enhancement of the resolution of terahertz images using deconvolution," in *Proc. SPIE 9856, Terahertz Physics, Devices, and Systems X: Advanced Applications in Industry and Defense, 98560N*, 2016, p. 98560N.
[14] K. Ahi and M. Anwar, "Embedding and Fabrication of Authentication Signatures by Robust Engineered Nanostructures," in *Diminishing Manufacturing Sources and Material Shortages (DMSMS)*, 2015.
[15] K. Ahi and M. Anwar, "A Novel Approach for Extracting Embedded Authentication Engineered NanoSignatures," in *Diminishing Manufacturing Sources and Material Shortages (DMSMS)*.
[16] K. Ahi, N. Asadizanjani, S. Shahbazmohamadi, M. Tehranipoor, and M. Anwar, "Terahertz characterization of electronic components and comparison of terahertz imaging with x-ray imaging techniques," in *Proc. SPIE 9483, Terahertz Physics, Devices, and Systems IX: Advanced Applications in Industry and Defense, 94830K*, 2015, p. 94830K.
[17] K. Ahi, N. Asadizanjani, M. Tehranipoor, and M. Anwar, "Authentication of electronic components by time domain THz Techniques," in *Connecticut Symposium on Microelectronics & Optoelectronics*, 2015.
[18] K. Ahi, N. Asadizanjani, S. Shahbazmohamadi, M. Tehranipoor, and M. Anwar, "THZ Techniques: A Promising Platform for Authentication of Electronic Components," in *CHASE Conference on Trustworthy Systems and Supply Chain Assurance*, 2015.
[19] K. Ahi, A. Rivera, A. Mazady, and M. Anwar, "Embedding Complex Nano-Signatures for Counterfeit Prevention in Electronic Components," in *CHASE Conference on Trustworthy Systems and Supply Chain Assurance*, 2015.
[20] K. Ahi, A. Rivera, A. Mazady, and M. Anwar, "Authentication of electronic components using embedded nano-signatures," in *Connecticut Symposium on Microelectronics & Optoelectronics*, 2015.
[21] K. Ahi, A. Mazady, A. Rivera, M. Tehranipoor, and M. Anwar, "Multi-level Authentication Platform Using Electronic Nano-Signatures," in *2nd International Conference and Exhibition on Lasers, Optics & Photonics*, 2014.
[22] W. Zhou and a C. Bovik, "A universal image quality index," *Signal Process. Lett. IEEE*, vol. 9, no. 3, pp. 81–84, 2002.
[23] Z. Wang, A. C. Bovik, H. R. Sheikh, and E. P. Simoncelli, "Image quality assessment: From error visibility to structural similarity," *IEEE Trans. Image Process.*, vol. 13, no. 4, pp. 600–612, 2004.
[24] Z. Wang and A. C. Bovik, "Mean Squared Error: Love It or Leave It?," *IEEE Signal Process. Mag.*, vol. 26, no. January, pp. 98–117, 2009.

Progression of Strain Relaxation in Linearly-Graded GaAs$_{1-y}$P$_y$/GaAs (001) Epitaxial Layers Approximated by a Finite Number of Sublayers

Tedi Kujofsa
Electrical and Computer Engineering Department,
University of Connecticut,
371 Fairfield Way, Unit 4157,
Storrs, CT 06269-4157, USA
tedi.kujofsa@gmail.com

John E. Ayers
Electrical and Computer Engineering Department,
University of Connecticut,
371 Fairfield Way, Unit 4157,
Storrs, CT 06269-4157, USA
john.ayers@uconn.edu

We have investigated the residual in-plane strain and width of the surface misfit dislocation free zone in linearly-graded GaAs$_{1-y}$P$_y$ metamorphic buffer layers as approximated by a finite number of sublayers. For this purpose we have developed an electric circuit model approach for the equilibrium analysis of these structures, in which each sublayer may be represented by an analogous configuration involving a current source, a resistor, a voltage source, and an ideal diode. The resulting node voltages in the analogous electric circuit correspond to the equilibrium strains in the original epitaxial structure. Utilizing this new approach, we show that the residual surface strain in linearly-graded epitaxial structures increases monotonically with grading coefficient as well as the number of sublayers, and is strongly dependent on the width of the misfit dislocation free zone, which diminishes with an increasing grading coefficient.

Keywords: electrical circuit model; relaxation; metamorphic buffer layers; strain; step-grading; linearly-graded; grading coefficient.

1. Introduction

The design of electronic and optical devices [1,2,3] often requires the use of metamorphic buffer layers to accommodate the misfit strain associated with the growth of mismatched materials. Understanding the equilibrium and kinetically-limited lattice relaxation mechanisms as well as the dislocation dynamics has important implications in the optimization of these device structures for better performance and reliability. Although several models [4-15] have been established for understanding equilibrium lattice relaxation, such approaches use specialized code, are computationally intense, and furthermore do not lend themselves to an intuitive understanding necessary for innovative structure design. Therefore, in this work we present an analogous electrical circuit model to understand equilibrium lattice relaxation in linearly-graded (LG) GaAsP on GaAs (001) epitaxial layers. Here we have approximated the linearly-graded material by a finite number of sublayers and show the effect of the number of sublayers on the progression of the residual strain characteristics. Furthermore, we have revisited the distribution of the in-plane strain as a function of the distance from the interface and give an improved analytical expression which is a direct product of the electrical circuit model. In the previously developed models, the main assumption was that misfit dislocation density is sufficient to completely relax the lattice mismatch in the dislocated region whereas in this work we account for the small residual strain in the dislocated material. As a consequence, the strain profile includes two departures from the model of Tersoff [13], which are a modification of the surface residual strain and a change in the width of the surface misfit dislocation free zone (MDFZ).

2. Physical Model of Equilibrium Lattice Relaxation

Matthews and Blakeslee [4] determined the equilibrium strain of a single epitaxial layer with uniform composition by considering a force-balance model in which the glide force on a grown-in dislocation is equated to the opposing line tension. The equilibrium strain may also be found by minimizing the sum of the strain energy and the line energy of misfit dislocations [5]. In a metamorphic epitaxial layer, assumed to have the diamond or zinc blende structure with (001) orientation and uniform composition (as shown in Fig. 1(a)), and which contains misfit dislocations, the in-plane strain ε is given by

$$\varepsilon = f - \frac{f}{|f|}\rho b \sin\alpha \sin\phi, \qquad (1)$$

where f is the lattice mismatch, $f \equiv (a_s - a_e)/a_e$, a_s and a_e are the relaxed lattice constant of the substrate and epitaxial crystal, respectively, the term $f/|f|$ accounts for the sign of the mismatch, ρ is the linear misfit dislocation density at the mismatched interface, b is the length of the Burgers vector, α is the angle between the Burgers vector and the line vector for the dislocation and ϕ is the angle between the glide plane and interface. The lattice mismatch of the substrate is defined as $f_0 \equiv 0$. In a metamorphic epitaxial layer with thickness h and an in-plane strain ε, the strain energy per unit area is

$$E_\varepsilon = Yh\varepsilon^2, \qquad (2)$$

where Y is the biaxial modulus, $Y = C_{11} + C_{12} - 2C_{12}^2/C_{11} = 2G(1+v)/(1-v)$, C_{11} and C_{12} are the elastic stiffness constants, G is the shear modulus and v is the Poisson ratio. The areal line energy of dislocations using a mean-field approach without including dislocation-dislocation interactions and assuming two orthogonal networks with equal cross-sectional density is

$$E_d = \rho \frac{Gb^2(1-\nu\cos^2\alpha)}{2\pi(1-\nu)}\left[\ln\left(\frac{h}{b}\right)+1\right]$$
$$= (f-\varepsilon)\frac{f}{|f|}\frac{Gb(1-\nu\cos^2\alpha)}{2\pi(1-\nu)\sin\alpha\sin\phi}\left[\ln\left(\frac{h}{b}\right)+1\right] \quad (3)$$

The equilibrium configuration is found by minimizing the sum of the dislocation line energy and the strain energy:

$$\frac{\partial E}{\partial \varepsilon} = \frac{\partial(E_\varepsilon + E_d)}{\partial \varepsilon} = 2Yh\varepsilon - \frac{f}{|f|}\frac{Gb(1-\nu\cos^2\alpha)}{2\pi(1-\nu)\sin\alpha\sin\phi}\left[\ln\left(\frac{h}{b}\right)+1\right] = 0. \quad (4)$$

The solution, accounting for the possibility of pseudomorphic growth, is

$$\varepsilon(h) = \begin{cases} f, & h \le h_c \\ \dfrac{f}{|f|}\dfrac{b(1-\nu\cos^2\alpha)}{8\pi h(1+\nu)\sin\alpha\sin\phi}\left[\ln\left(\dfrac{h}{b}\right)+1\right], & h > h_c \end{cases}, \quad (5)$$

where h_c is the critical layer thickness at which it becomes energetically favorable to introduce misfit dislocations. Below h_c, the in-plane strain is equal to the lattice mismatch.

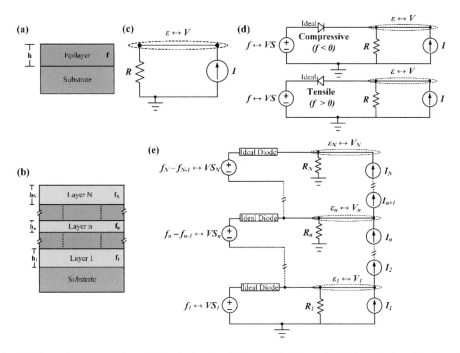

Fig. 1. Schematic representation of (a) a single epitaxial layer and (b) an epitaxial layer broken down into N arbitrary sublayers. (c) Simple resistive circuit consisting of a resistor and independent current source. (d-e) The equivalent electrical circuit for (d) a single epitaxial layer with both compressive and tensile cases and (e) an epitaxial layer broken down into N sublayers.

The equilibrium model can be generalized to any multilayered or compositionally-graded structure by considering a stack of N disparate layers as shown in Fig. 1(b). (The substrate is considered layer zero in this work.) The strain ε_n in the nth sublayer of the stack may be determined by accounting for the misfit dislocation densities concentrated in that layer and those below it:

$$\varepsilon_n = f_n - \frac{f_n - f_{n-1}}{|f_n - f_{n-1}|} \sum_{j=1}^{n} \rho_n b_n \sin\alpha \sin\phi . \tag{6}$$

The energy E of the structure illustrated in Fig. 1(b) is found by summing the contributions of the areal strain and dislocation line energies of each sublayer,

$$E = \sum_{j=1}^{N} E_{d,j} + E_{\varepsilon,j} . \tag{7}$$

To determine the equilibrium strain, we differentiate the energy with respect to the in-plane strain at each sublayer, set each partial derivative to zero, and concurrently solve the system of N equations:

$$0 = \frac{\partial E}{\partial \varepsilon_n} = 2Y_N h_N \varepsilon_N - \frac{f_N - f_{N-1}}{|f_N - f_{N-1}|} \frac{G_N b_N \left(1 - v_N \cos^2\alpha\right)}{2\pi(1-v_N)\sin\alpha\sin\phi} \left[\ln\left(\frac{h_N}{b_N}\right) + 1 \right], \quad n = N$$

$$0 = \frac{\partial E}{\partial \varepsilon_n} = \begin{pmatrix} 2Y_n h_n \varepsilon_n - \frac{f_n - f_{n-1}}{|f_n - f_{n-1}|} \frac{G_n b_n \left(1 - v_n \cos^2\alpha\right)}{2\pi(1-v_n)\sin\alpha\sin\phi} \left[\ln\left(\sum_{j=n}^{N} \frac{h_j}{b_n}\right) + 1 \right] \\ + \frac{f_{n+1} - f_n}{|f_{n+1} - f_n|} \frac{G_{n+1} b_{n+1} \left(1 - v_{n+1} \cos^2\alpha\right)}{2\pi(1-v_{n+1})\sin\alpha\sin\phi} \left[\ln\left(\sum_{j=n+1}^{N} \frac{h_j}{b_{n+1}}\right) + 1 \right] \end{pmatrix}, \quad 1 \leq n < N \tag{8}$$

Essentially, the analysis provided above is the basis for the development of the Bertoli et al. [6] model, which utilizes an ad hoc numerical approach to minimize the energy for an arbitrary multilayered heterostructure. The approach is generally applicable, and compositionally-graded layers may be represented by staircase profiles with arbitrary precision.

3. Equilibrium Circuit Model for Equilibrium Lattice Relaxation

The development of an electric circuit model stems from the fact that Eq. (4) resembles the node voltage expression [7] for the top node of the simple electrical circuit shown in Fig. 2(a):

$$0 = 2Yh\varepsilon - \frac{f}{|f|} \frac{Gb(1-v\cos^2\alpha)}{2\pi(1-v)\sin\alpha\sin\phi} \left[\ln\left(\frac{h}{b}\right) + 1 \right] \quad \leftrightarrow \quad \frac{V}{R} - I = 0 . \tag{9}$$

The node voltage equation is derived from Kirchhoff's current law, which states that the algebraic sum of the currents flowing away from a node must equal zero[*]. The symbol $\leftrightarrow$ implies that quantities or relationships on either side of the arrow are analogous though they generally possess different units. If we consider that the partial derivatives are analogous to electrical currents, comparison of the two forms of Eq. (9) reveals that the equilibrium strain is analogous to the node voltage[†],

$$\varepsilon \leftrightarrow V, \tag{10}$$

the factor multiplying the strain is analogous to a conductance (reciprocal of resistance),

$$2Yh \leftrightarrow 1/R, \tag{11}$$

and the subtracted term is analogous to an independent current source entering the top node of the circuit,

$$\frac{f}{|f|} \frac{Gb(1-v\cos^2\alpha)}{2\pi(1-v)\sin\alpha\sin\phi}\left[\ln\left(\frac{h}{b}\right)+1\right] \leftrightarrow I. \tag{12}$$

In an electrical circuit, each node voltage is defined with respect to the ground reference node which is analogous to the substrate in an epitaxial structure. To account for the possibility of pseudomorphic epitaxy, we can include an independent voltage source and an ideal diode in the circuit, which together form a clipping circuit as shown in Fig. 2(b). The ideal diode acts as a switch that is conductive only when an epitaxial layer is coherently grown and no misfit dislocation network is introduced at the interface with the underlying layer. For the case of a single uniform layer, the numerical value of the voltage source VS is equal to the coherency strain in the layer,

$$f \leftrightarrow VS. \tag{13}$$

To properly account for the sign of the lattice mismatch (tensile or compressive), the ideal diode must always face toward *the true positive terminal* of the independent voltage source (Fig. 2(b) illustrates both cases). Therefore, in terms of the electrical circuit model, the two analogous forms of the solution for the node voltage (or the equivalent equilibrium strain) are given as

$$V = \begin{cases} VS, & h \leq h_c \\ I \cdot R, & h > h_c \end{cases} \leftrightarrow \varepsilon(h) = \begin{cases} f, & h \leq h_c \\ \dfrac{f}{|f|} \dfrac{b(1-v\cos^2\alpha)}{8\pi h(1+v)\sin\alpha\sin\phi}\left[\ln\left(\dfrac{h}{b}\right)+1\right], & h > h_c \end{cases}. \tag{14}$$

We can extend the electrical circuit model described above to the generalized heterostructure of Fig. 1(b) utilizing a similar approach such that Eq. (8) resembles the node voltage expressions for N essential nodes[‡] [7]. Therefore, we can consider the consecutive

---

[*] In this analysis, an electrical current which enters the node is considered negative.
[†] The choice of analogies used here is not unique, but is selected for its convenience. Other consistent choices of analogies will yield the same solution.
[‡] An essential node is defined as a node connected to more than two circuit elements.

stacking of the electrical circuit blocks given in Fig. 2(b) to obtain an equivalent circuit that describes an N layered heterostructure (Fig. 2(c)). In the general case, the partial derivatives shown in Eq. (8) are given in terms of their analogous electrical circuit components as

$$0 = 2Y_N h_N \varepsilon_N - \frac{f_N - f_{N-1}}{|f_N - f_{N-1}|} \frac{G_N b_N (1 - v_N \cos^2 \alpha)}{2\pi (1 - v_N) \sin \alpha \sin \phi} \left[\ln\left(\frac{h_N}{b_N}\right) + 1 \right] \quad \leftrightarrow \quad \frac{V_N}{R_N} - I_N = 0, \quad n = N$$

$$0 = \begin{pmatrix} 2Y_n h_n \varepsilon_n - \dfrac{f_n - f_{n-1}}{|f_n - f_{n-1}|} \dfrac{G_n b_n (1 - v_n \cos^2 \alpha)}{2\pi (1 - v_n) \sin \alpha \sin \phi} \left[\ln\left(\displaystyle\sum_{j=n}^{N} \dfrac{h_j}{b_n}\right) + 1 \right] \\ + \dfrac{f_{n+1} - f_n}{|f_{n+1} - f_n|} \dfrac{G_{n+1} b_{n+1} (1 - v_{n+1} \cos^2 \alpha)}{2\pi (1 - v_{n+1}) \sin \alpha \sin \phi} \left[\ln\left(\displaystyle\sum_{j=n+1}^{N} \dfrac{h_j}{b_{n+1}}\right) + 1 \right] \end{pmatrix} \quad \leftrightarrow \quad \frac{V_n}{R_n} - I_n + I_{n+1} = 0, \quad 1 \le n < N \qquad (15)$$

It can be shown that the numerical value of the voltage at each essential node is equivalent to the equilibrium strain of the corresponding sublayer,

$$\varepsilon_n \leftrightarrow V_n. \qquad (16)$$

In this extended analogy, the nth sublayer may be modeled by a subcircuit in which:

$$R_n \leftrightarrow \frac{1}{2Y_n h_n}, \quad 1 \le n \le N, \qquad (17)$$

$$I_n \leftrightarrow \frac{f_n - f_{n-1}}{|f_n - f_{n-1}|} \frac{G_n b_n (1 - v_n \cos^2 \alpha)}{2\pi (1 - v_n) \sin \alpha \sin \phi} \left[\ln\left(\sum_{j=n}^{N} \frac{h_j}{b_n}\right) + 1 \right], \quad 1 \le n \le N, \qquad (18)$$

and the diode-connected independent voltage source is determined by the difference in mismatch of the two adjacent layers:

$$VS_n \leftrightarrow f_n - f_{n-1}, \quad 1 \le n \le N. \qquad (19)$$

In the case in which each sublayer contains misfit dislocations, none of the diodes conduct, and the in-plane strain (node voltage) at the nth sublayer is determined by

$$\varepsilon_n \leftrightarrow V_n = \begin{cases} R_n \cdot (I_n - I_{n+1}), & 1 \le n < N \\ R_n \cdot I_n, & n = N \end{cases}. \qquad (20)$$

However, the growth of metamorphic epitaxial layers containing compositional grading may sometimes result in the creation of misfit dislocation free zones (MDFZ) [4,6,8,9,10,11,12,13,14,15]. If the epitaxial structure is incoherent as a whole but some of the sublayers are coherently-grown with respect to the ones below, the presence of a MDFZ may be likened to the formation of a supernode[§] [7] in the analogous electrical circuit model. The existence of the supernode modifies the node voltage equations for the nodes involved, and therefore the resulting voltages. If the supernode is bounded inclusively by

---

[§] A supernode in electrical circuit theory refers to the situation in which two essential nodes are separated by an independent voltage source.

sublayers σ and ω, then the equilibrium strain (node voltage) in the bottom layer of the supernode is given by

$$\varepsilon_\sigma \leftrightarrow V_\sigma = \left[(I_\sigma - I_{\omega+1}) - \sum_{j=\sigma}^{\omega} \frac{\sum_{i=1}^{j} V_{Si} - \sum_{i=1}^{\sigma} V_{Si}}{R_j} \right] R_{SN}, \qquad (21)$$

where the equivalent parallel resistance of the supernode R_{SN} is defined as the equivalent resistance for a configuration of resistors in parallel, $R_{SN} = R_\sigma \|\ldots\| R_\omega$, and is given by

$$R_{SN} = \left(\sum_{j=\sigma}^{\omega} \frac{1}{R_j} \right)^{-1}. \qquad (22)$$

The in-plane strain (node voltage) at each sublayer of the *supernode* is then determined by adding the appropriate sum of independent voltage sources to the voltage at the bottom of the supernode. In other words, the node voltage (or the equivalent in-plane strain) of each sublayer (not including the bottom layer of the supernode) of the *supernode* is determined from

$$\varepsilon_i = \varepsilon_\sigma + \sum_{j=\sigma+1}^{i} f_j - f_{j-1} \leftrightarrow V_i = V_\sigma + \sum_{j=\sigma+1}^{i} VS_j, \quad \sigma < i \leq \omega, \qquad (23)$$

4. Results and Discussion

In this work, we investigated the grading coefficient dependence of the equilibrium strain relaxation by varying the number of sublayers, the ending lattice mismatch or the total thickness of the epitaxial layer. In this work, we have defined the average grading coefficient as $C_f = f_h / h$ where f_h is the lattice mismatch of the surface and h is the total epilayer thickness. In a linearly-graded layer, the slope of the mismatch versus thickness characteristic is equal to the average grading coefficient. For a step-graded structure, the change in mismatch at each step is $\Delta f = C_f / N$. Figure 2(a) shows a compositionally uniform, 150 nm thick layer of $GaAs_{0.7}P_{0.3}$ on GaAs (001) and the equivalent electrical circuit. The material parameters used here are summarized in Table 1. Using the Matthews and Blakeslee model, the equilibrium in-plane strain is $\varepsilon = 0.1024\%$ and this is equal to the numerical value of the node voltage provided by the equivalent electrical circuit. Whereas the strain is unitless, the node voltage is in Volts (V). Figure 2(b) considers a step-graded $GaAs_{1-y}P_y$ epitaxial layer with three 50 nm thick sublayers in which there are equal compositional changes from one layer to the next. The composition in the top sublayer is fixed at 30% phosphorus. Strain relaxation behavior in a step-graded structure is dictated by the behavior of the individual uniform-composition sublayers, each of which exhibits a concentration of misfit dislocations at the interface and a uniform residual strain in the remaining thickness. Some interfaces may be absent of misfit dislocations, so the width of the surface MDFZ may be an integral multiple of the step thickness. The results of Fig. 2(b) show that misfit dislocation networks are present at all interfaces and therefore the width of the surface MDFZ is fixed to that of the top sublayer thickness of 50 nm.

However, when increasing the number of sublayers and therefore utilizing a lower compositional change at each interface as shown in Fig. 2(c), it becomes apparent that the width of the surface MDFZ increases. Figure 2(c) shows that the fifth sublayer is coherent with respect to the one below and therefore the surface MDFZ in this case is twice the sublayer thickness, or 60 nm. Although the average strain is comparable among the three structures illustrated here, the surface strain increases with the average grading coefficient or the number of sublayers.

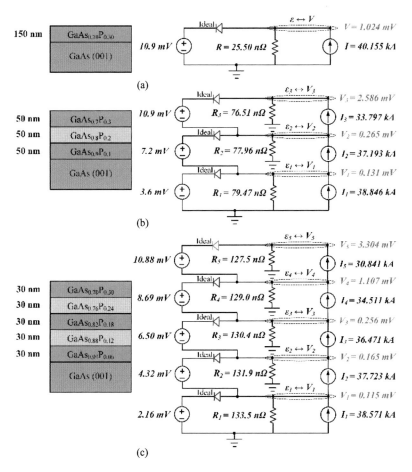

Fig. 2. Schematic representation of step-graded epitaxial layers and the equivalent electrical circuit with (a) a single, (b) three and (c) five sublayers.

Table 1. Material Properties for GaP, GaAs and the alloy $GaAs_{1-y}P_y$.

Parameter	GaP	$GaAs_{1-y}P_y$	GaAs
a (nm)	0.54505	0.56534 – y(0.02029)	0.56534
C11 (GPa)	140.5	118.4 + y(22.1)	118.4
C12 (GPa)	62.03	53.7 + y(8.33)	53.7

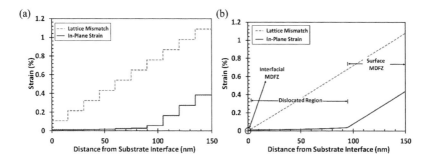

Fig. 3. Lattice mismatch and in-plane strain as a function of the distance from the interface for a step-graded with 10 sublayers and (b) a linearly-graded epilayer of GaAs$_{0.7}$P$_{0.3}$ on GaAs (001) substrate.

These characteristic behaviors are more evident in Fig. 3, which shows the lattice mismatch and in-plane strain as a function of the distance from the interface for a step-graded layer with 10 sublayers as well as a linearly-graded epilayer approximated with 200 sublayers. In both cases, the ending composition is fixed at 30% phosphorus and the total epilayer thickness is 150 nm; however for the linearly-graded structure, the phosphorus composition at the substrate interface is fixed at zero to match the substrate. The results of Fig. 3 show a monotonic increase in the surface MDFZ width with the number of sublayers. For the case of linear-grading the surface MDFZ width is 55 nm whereas in the step-graded layer it is constrained to 45 nm (three times the sublayer thickness). In addition, because of the low grading coefficient used in LG epitaxial layers, the misfit dislocations are introduced at a finite distance from the substrate interface, which results in the formation of an interfacial MDFZ as pointed out in Fig. 3(b). The interfacial MDFZ is shown in the circled region of Fig. 3(b), but is difficult to see with the scale of this figure because its thickness is only a few nanometers. Although increasing the number of sublayers does not result in any noticeable changes in the average residual strain, the results of Figs. 2 and 3 show that the surface in-plane strain changes significantly from 0.102% for a single layer to 0.434% for 200 sublayers.

The characteristics of Figs. 4 and 5 investigate the effect of the grading coefficient as well as the number of sublayers on the surface strain. Figure 4 illustrates that the surface in-plane strain as a function of the number of sublayers (Fig. 4(a)) and its reciprocal (Fig. 4(b)) with ending lattice mismatch as a parameter for a 150 nm thick GaAs$_{1-y}$P$_y$ epitaxial layer on GaAs (001). In these structures, the phosphorus composition at the top sublayer is fixed at 30%, 40% and 50% respectively corresponding to an ending lattice mismatch of 1.08%, 1.45% and 1.82%. The structures associated with a particular symbol in both Figs. 4 and 5 contain identical grading coefficients but we can compare them in terms of the number of sublayers. Figure 5 displays similar features however, the adjusted parameter is the epitaxial layer thickness; for the structures shown in Fig. 5, the phosphorus composition at the top sublayer is fixed at 40% and the epitaxial layer thickness is varied to 150, 300 and 600 nm. The characteristics of Figs. 4 and 5 demonstrate a sublinear and monotonically increasing surface in-plane strain with the number of sublayers. More specifically, there is an increase in the residual surface strain when there is a combination of an (i) increase in the ending lattice mismatch, (ii) a decrease in the total epitaxial layer thickness and (iii) an increase in the total number of sublayers.

The slight departures from smoothness (especially in the region where N ranges from 8 to 16 sublayers) of the surface in-plane strain observed in Figs. 4 and 5 could be explained with the aid of Fig. 6. Previously, we mentioned that in step-graded layers the width of the surface MDFZ is a multiple of the sublayer number and therefore the actual width is determined from the product of the number of coherent interfaces near the surface and the width of each sublayer. Figure 6 shows the width of the surface MDFZ (left vertical axis) and the number of coherent interfaces (right vertical axis) as a function of the number

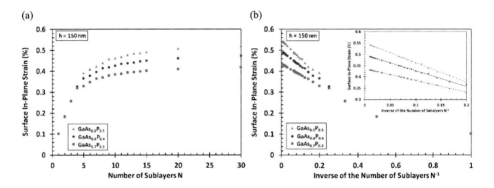

Fig. 4. Surface in plane strain as a function of (a) the number of sublayers and (b) its reciprocal for a linearly-graded layer approximated with a finite number of sublayers and the ending lattice mismatch as a parameter. The epitaxial layer thickness in these structures is fixed at 150 nm. The inset of part (b) shows a subset of the data shown on Fig. 4(b) which are associated with a higher number of sublayers. The axis labels for the inset figure are the same as those of Fig. 4(b).

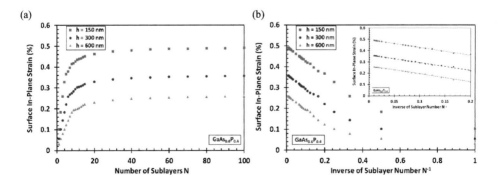

Fig. 5. Surface in plane strain as a function of (a) the number of sublayers and (b) its reciprocal for a linearly-graded layer approximated with a finite number of sublayers and the epitaxial layer thickness as a parameter. The phosphorus composition at the top sublayer is fixed at 40 corresponding to a lattice mismatch of 1.45%. The inset of part (b) shows a subset of the data shown on Fig. 5(b) which are associated with a higher number of sublayers. The axis labels for the inset figure are the same as those of Fig. 5(b).

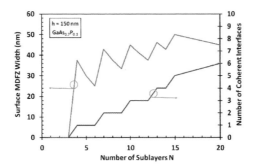

Fig. 6. (a) Surface MDFZ width and (b) number of coherent interfaces near the surface as a function of the number of sublayers for a 150 nm thick GaAs$_{0.7}$P$_{0.3}$/GaAs (001).

of sublayers for a 150 nm thick GaAs$_{0.7}$P$_{0.3}$ on a GaAs (001) substrate. Due to the approximation of the linear-grading scheme with a finite number of sublayers, it can be seen that although there is an increase in the number of sublayers, at a small number of sublayers (N < 20), there is a non-monotonic variation of the width of the surface MDFZ. In addition, the results of Fig. 6 show that there are structures in which the number of coherent interfaces remains the same when increasing the number of sublayers slightly as is the case for N = 4 → 6, 7 → 9, 10 → 12 and 13 → 14. Even though the residual strain increases with the number of sublayers, for small number of sublayers the competing mechanisms of the surface MDFZ width and the incoherent substrate lead observed departures in the smoothness of residual strain characteristic.

In the limiting case, we can consider a linearly-graded epitaxial layer with a large number of sublayers, and the characteristics shown above become more apparent. Figure 7(a) shows the width of the surface misfit dislocation free zone and the surface in plane strain as a function of average grading coefficient. The width of the surface MDFZ decreases sublinearly and monotonically with increasing grading coefficient and this is expected on the basis that higher mismatch requires the introduction of more misfit dislocations which leads to the extension of the dislocations close to the surface and therefore in the diminishment of the surface MDFZ. It is interesting to note that the results of Fig. 7(b) demonstrate that surface in-plane strain is strongly dependent on the width of the surface MDFZ. The novelty of the electrical circuit model is that it enables a complete understanding of the strain profile in linearly-graded structures. Although, Tersoff [13,14] and Fitzgerald et al. [15] have developed models for the distribution of the in-plane strain as a function of the distance from the substrate interface, these models were approximate because they neglected the interfacial MDFZ as well as the finite in-plane strain in the dislocated region.

If the edges of the interfacial and surface MDFZs are located at distances of z_1 and z_2 from the substrate interface, and therefore the misfit dislocation density is concentrated in the middle region ($z_1 \leq z \leq z_2$), then the residual strain can be analytically modeled as follows: In the interfacial MDFZ, the absence of misfit dislocations indicates that the residual strain is equal to the lattice mismatch profile and therefore:

$$\varepsilon(z) = C_f z, \quad z \leq z_1. \tag{24}$$

In the dislocated region ($z_1 \leq z \leq z_2$), the residual strain is modeled by the electrical circuit model as

$$\varepsilon_n \leftrightarrow V_n = R_n \cdot (I_n - I_{n+1}) = \frac{(I_n - I_{n+1})}{2Y_n h_n}. \tag{25}$$

If we consider the limiting case where $h_n \to 0$, then it can be shown that

$$\lim_{h_n \to 0} \{\varepsilon_n\} = \varepsilon(z) = \frac{A}{h-z}, \quad 0 \leq z \leq z_1, \tag{26}$$

where

$$A = -\frac{f'(z)}{|f'(z)|} \frac{b(1 - \nu \cos^2 \alpha)}{8\pi(1+\nu)\sin\alpha \sin\phi}. \tag{27}$$

where $f'(z)$ is the first order derivative of the lattice mismatch profile. The characteristic of Eq. (27) describes the residual strain in structures where the misfit dislocation region extends all the way to the substrate interface, however in linearly-graded epitaxial layers,

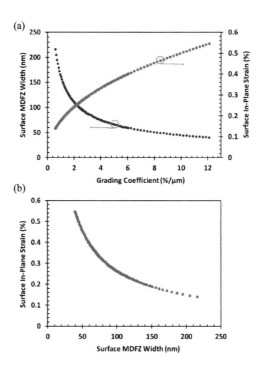

Fig. 7. (a) Surface MDFZ width and the in-plane strain as a function of the grading coefficient for GaAsP/GaAs (001). (b) Surface in-plane strain as a function of the surface MDFZ width.

the presence of the interfacial MDFZ leads to the adjustment of the strain profile. In this case, the residual strain in the dislocated region is modeled by

$$\varepsilon(z) = \frac{A}{h-z} - \frac{A}{h} + C_f z_1, \quad z_1 < z \le z_2. \tag{28}$$

The second and third terms in the equation above represents adjustments to account for the strain at the top of the interfacial MDFZ. By a similar analysis, in the surface MDFZ, the absence of misfit dislocations implies that the residual strain is proportional to the lattice mismatch and therefore the strain in this region is given by

$$\varepsilon(z) = C_f(z - z_2) + \frac{A}{h - z_2} - \frac{A}{h} + C_f z_1, \quad z_2 < z \le h \tag{29}$$

Therefore, according to the above model, the equilibrium strain profile in the linearly-graded layer is given by

$$\varepsilon(z) = \begin{cases} C_f z & z \le z_1; \\ A \dfrac{z}{h(h-z)} + C_f z_1 & z_1 < z \le z_2; \text{ and} \\ C_f(z - z_2) + A \dfrac{z_2}{h(h - z_2)} + C_f z_1 & z_2 < z \le h. \end{cases} \tag{30}$$

The condition for the surface MDFZ boundary z_2 is given by

$$\int_{z_2}^{h} \varepsilon(z) dz = \frac{E_d}{2b'Y} \leftrightarrow \int_{z_2}^{h} \left(C_f(z - z_2) + \varepsilon(z_2) \right) dz = A \left[\ln\left(\frac{h - z_2}{b} \right) + 1 \right]. \tag{31}$$

Solving the expression above and recognizing that the width of the surface MDFZ is $W_{MDFZ} = h - z_2$, yields the surface in-plane strain characteristic shown in Fig. 7(b) to be accurately modeled by

$$\varepsilon_S = \frac{A}{W_{MDFZ}} \ln\left(\frac{W_{MDFZ}}{b} \right) + \frac{A}{h} - C_f z_1. \tag{32}$$

The sum of the second and third terms yields a small contribution to the equation since the boundary for the interfacial misfit dislocation free zone z_1 is very small in these structures, however, its value could be found by a similar approach where

$$\int_{0}^{z_1} \varepsilon(z) dz = \frac{E_d}{2b'Y} \leftrightarrow \int_{0}^{z_1} \left(C_f z \right) dz = A \left[\ln\left(\frac{h - z_1}{b} \right) + 1 \right]. \tag{33}$$

Solving, the expression above results in transcendental expression similar to the Matthews and Blakeslee critical layer thickness equation,

$$z_1 = \sqrt{\frac{2A}{C_f} \left[\ln\left(\frac{h - z_1}{b} \right) + 1 \right]}. \tag{34}$$

Therefore, Eq. (33) is modified accordingly to

$$\varepsilon_S = \frac{A}{W_{MDFZ}} \ln\left(\frac{W_{MDFZ}}{b}\right) + \frac{A}{h} - \sqrt{2AC_f\left[\ln\left(\frac{h-z_1}{b}\right)+1\right]}. \tag{35}$$

5. Conclusion

We have investigated equilibrium lattice relaxation in metamorphic in $GaAs_{1-y}P_y/GaAs$ (001) heterostructures involving linear grading in composition by utilizing an analogous electrical circuit model. We have approximated the linear grading scheme using a finite number of sublayers and have explored its effect on the surface strain and the width of the misfit dislocation free zone. There are two key conclusions to this paper: first, the surface in-plane strain increase with greater grading coefficient and second, the value of the surface strain is strongly dependent on the width of the surface MDFZ which diminishes at higher mismatch.

References

[1] K. Streubel, N. Linder, R. Wirth, and A. Jaeger, IEEE J. Sel. Top. Quant. Electron., 8, 321 (2002).
[2] P. Lever, H. H. Tan, and C. Jagadish, J. Appl. Phys., 95, 5710 (2004).
[3] S. Mendach, C. M. Hu, Ch. Heyn, S. Schnull, H. P. Oepen, R. Anton, and W. Hansen, Physica E, 13, 1204 (2002).
[4] J. W. Matthews and A. E. Blakeslee, J. Cryst. Growth 27, 118 (1974).
[5] J. W. Matthews, *Epitaxial Growth, Part B* (Academic Press, New York, NY, 1975).
[6] B. Bertoli, E. N. Suarez, J. E. Ayers, and F. C. Jain, J. Appl. Phys., 106, 073519 (2009).
[7] J. W. Nielsen and Susan Riedel, *Electric Circuits 8th Edition* (Pearson Prentice Hall, Upper Saddle River, NJ, 2008.)
[8] T. Kujofsa and J. E. Ayers, J. Vac. Sci. Technol. B, (32), 031205 (2014).
[9] T. Kujofsa, A. Antony, S. Xhurxhi, F. Obst, D. Sidoti, B. Bertoli, S. Cheruku, J. P. Correa, P. B. Rago, E. N. Suarez, F. C. Jain, and J. E. Ayers, J. Electron. Mat., 42, 3408 (2013).
[10] T. Kujofsa and J. E. Ayers, J. Electron. Mat., 45, 2831 (2016).
[11] T. Kujofsa and J. E. Ayers, J. Electron. Mat., 43, 2993 (2014).
[12] J. E. Ayers, *Heteroepitaxy of Semiconductors: Theory, Growth, and Characterization* (CRC Press, Boca Raton, FL, 2007).
[13] J. Tersoff, Appl. Phys. Lett. 62, 693 (1993).
[14] J. Tersoff, Appl. Phys. Lett. 64, 2748 (1994).
[15] E. A. Fitzgerald, Y.-H. Xie, D. Monroe, P. J. Silverman, J. M. Kuo, A. R. Kortan, F. A. Thiel and B. E. Weir, J. Vac. Sci. Technol. B 10, 1807 (1992).

Carbon Nanotubes, Nanofibers and Nanospikes for Electrochemical Sensing: A Review

Aysha S. Shanta[*], Khandakar A. Al Mamun, Syed K. Islam and Nicole McFarlane

Department of Electrical Engineering and Computer Science,
University of Tennessee, Knoxville, TN 37996, USA
[*]*ashanta1@vols.utk.edu*

Dale K. Hensley

Center for Nanophase Materials Sciences,
Oak Ridge National Laboratory, Oak Ridge, TN 37831, USA

The structural and material properties of carbon based sensors have spurred their use in biosensing applications. Carbon electrodes are advantageous for electrochemical sensors due to their increased electroactive surface areas, enhanced electron transfer, and increased adsorption of target molecules. The bonding properties of carbon allows it to form a variety of crystal structures. This paper performs a comparative review of carbon nanostructures for electrochemical sensing applications. The review specifically compares carbon nanotubes (CNT), carbon nanofibers (CNF), and carbon nanospikes (CNS). These carbon nanostructures possess defect sites and oxygen functional groups that aid in electron transfer and adsorption processes.

Keywords: biosensor; carbon nanotubes; carbon nanofibers; carbon nanospikes.

1. Introduction

A biosensor is defined as an analytical device and consists of a biological sensing element integrated with a transducer. The transducer converts the biological response into an electrical signal. Biosensors based on electrochemical methods are widely used and their high signal-to-noise ratio, high sensitivity, simplicity, and reduced response time have resulted in increasing demand for their use in the analysis of biological and environmental analytes. Carbon based nanomaterials are emerging as popular materials for the development of advanced sensor technologies. Development of an efficient biosensor technology generally requires the resolution of the following issues:

1. Development of biosensing interfaces so that the analytes can easily bond with the surface.
2. Development of an efficient transducer to convert the biosensing event to an equivalent electrical signal.

[*]Corresponding author.

3. Increasing the sensitivity and the selectivity of the biosensor.
4. Improving the response time of the sensor.

The material properties of carbon nanomaterials facilitate the resolution of these issues and allow the rapid development of biosensors technologies using this class of nanomaterials as front-end transducers.

Among the carbon nanomaterials, carbon nanotubes (CNT) have been used extensively in electrochemical sensors for detecting biomolecules. Recent research has focused on other carbon nanomaterials such as carbon nanohorns, nanofibers, and nanospikes[1-3]. Different forms of graphene are also being increasingly used as sensor electrodes[4,5]. This paper presents a comparative study of three types of carbon nanomaterials: nanotubes, nanofibers, and nanospikes. Carbon nanofibers (CNF) have a higher ratio of surface active groups, easier mass reproducibility, and low production cost making it a potential candidate for replacing carbon nanotubes[6]. Carbon nanospikes (CNS) are also being explored to replace both CNTs and CNFs due to their increased batch reproducibility and are considered to be a suitable solution to achieving CMOS integration[7-13]. Electrochemical sensing has excellent repeatability, accuracy and is less expensive compared to other detection methods. It is typically used to detect analytes which undergo a reduction/oxidation reaction such as glucose and dopamine. Table I displays the common properties and the biosensing applications of carbon nanostructures when used as electrochemical electrodes.

Table I. Properties and Biosensing Applications of Carbon Nanostructures.

Properties	Common Biological Analytes
• Submicron diameter sizes • High tensile strength and elastic modulus • High electric conductivity • Larger functionalized surface area • One-dimensional macrostructures • Facilitates electron transfer of electroactive analytes	• Dopamine and serotonin • Glucose • H_2O_2 • Acetylcholine • Alcohol • DNA • Cholesterol • Proteins, biomarkers and bacteria

This paper is organized as follows. Sections 2, 3, and 4 outline the basic properties of carbon nanostructures and Sec. 5 offers an overview of the fabrication process of each class of nanomaterial. Section 6 details the advantages and disadvantages of each electrode type as a function of electrode properties, CMOS integration, and electrochemical sensing. Finally, we provide a summary in Sec. 7.

2. Carbon Nanotubes (CNT)

The electrical and the mechanical properties of CNTs have led to its popularity in various scientific and technological applications. A wide variety of research has been conducted to utilize its applications as sensors, electrodes, hydrogen storage devices, and field

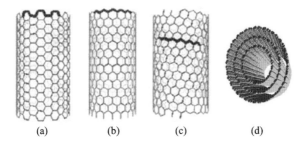

Fig. 1. Structural models of carbon nanotubes depending on the wrapping of graphene layers (a) single wall carbon nanotube armchair (b) single wall carbon nanotube: zig-zag (c) single wall carbon nanotube: chiral (d) multi wall carbon nanotube made up of three SWCNTs[18].

emission devices[14]. CNTs feature cylindrical tube-like shells of benzene-type hexagonal rings and possess high aspect ratio with diameters ranging from 10 to 50 nm and lengths up to a few micrometers. CNTs typically demonstrate metallic or semiconducting properties depending on their diameter and chirality. Single-walled CNTs (SWCNTs) are cylindrical structures of one atomic layer thick graphene sheets (Fig. 1). Multi-walled CNTs (MWCNTs) are graphene sheets rolled up in concentric circles[15-18].

3. Carbon Nanofibers (CNF)

Carbon nanofibers can be divided into platelet, tubular and herringbone structures according to the arrangement of graphene layers. The graphene layers of herringbone CNFs are inclined towards the axis of the fiber and adjusting of the angle of the graphene layer to the axis modulates the ratio of the basal atoms to the edge atoms. Vertically aligned carbon nanofibers, or VACNFs, have gained popularity because of the presence of numerous defect sites along their sidewalls. VACNFs are conical structures of stacked graphene layers with lengths in the order of microns and the diameter ranging from tens of nanometers to several hundred nanometers. Figure 2(a) shows the scanning electron microscope (SEM) image of a forest of VACNF which is used to determine the surface morphology of the fiber. Figure 2(b) shows the transmission electron microscope (TEM) of a single VACNF with crystalline walls at the center of the structure and broken walls at the surface of the structure with locations for electrochemical reactions throughout the structure[19-20]. VACNF can be grown as a single fiber or as a forest of fibers. Single

Fig. 2. (a) Scanning electron microscope image showing a forest of vertically aligned carbon nanofibers and, (b) transmission electron microscope image of a single vertically aligned carbon nanofiber[20].

nanofibers allow control over the spacing and the morphology of the VACNF electrodes and are desirable for high resolution sensing, while fiber forests can provide increased electrochemically active surface areas[21].

4. Carbon Nanospikes (CNS)

A thin layer of carbon nanomaterial coating on a conducting surface increases the number of defect sites leading to better adsorption of biological species on the surface of the electrodes. Deposition of catalyst during the growth process of carbon nanomaterials (such as VACNF) hinders the reproducibility of the resulting electrodes. Carbon nanospikes (CNS) contain spike-like structures with diameters on the order of nanometers[3]. This class of carbon nanomaterials demonstrate excellent electrochemical properties required for reduction/oxidation reactions without the need of a catalyst in the fabrication process. This allows substrates to be completely coated with carbon films and results in good electrical contact between the CNS and the metallic substrate. More importantly, batches of CNS can be fabricated with high reproducibility[22].

5. Electrode Fabrication

The growth procedure of carbon nanostructures relies on the property that filamentous carbon can be deposited on hot metal using catalytic deposition of gases containing carbon. Carbon can be formed by the interaction of hydrocarbons and metals such as nickel, iron and cobalt. In 1990, carbon nanotubes were grown by arc discharge synthesis[14]. The development of catalytic plasma enhanced chemical vapor deposition (c-PECVD) provided additional control over the carbon nanostructure fabrication process[23, 24]. It encouraged controlled deterministic growth resulting in user-controlled diameter, length, position, alignment, and chemical composition[25].

Arc discharge and laser ablation have been used to grow CNTs, requiring growth temperatures up to 4000 °C. A chamber filled with inert gas and the presence of a large current between two carbon electrodes generates carbon plasma. The amount of carbon deposited on the larger electrode depends on the chamber pressure and the nature of current flow[17]. An example of CNTs grown using this process is shown in Fig. 3. The growth temperature of carbon nanostructures using CVD ranges from 400 °C to 1000 °C.

Fig. 3. Carbon nanotube fabricated using two carbon electrodes[17].

The CVD process has been used to grow CNTs on patterned surfaces using metal catalysts for use as sensors, electronic devices and field emitters.

Carbon nanofibers exhibit either tip-growth mode or base-growth mode[14]. In the tip-growth mode, catalytic particles are detached from the substrate and are found at the tip of the carbon deposition. In the base-growth mode, the catalyst remains attached to the substrate and is found at the base of the growth. PECVD allows for lower wafer temperature and use of a plasma source creates vertically aligned structures. Nickel, cobalt or iron are commonly used catalysts in the fabrication process. The plasma generator decomposes a carbon-containing gas, such as acetylene and the electric field present in the chamber enhances the vertical growth[18]. The substrate temperature is maintained between 550 °C to 700 °C under continuous gas flow. The locations of the fiber growth is determined by the catalyst locations and hence the term deterministic fabrication. The formation of the nanostructures is greatly dependent on the plasma power, the ratio of hydrocarbon to etchant gases, the flow rate, the growth time, and the catalyst size. The alignment of the growth is controlled by the electric field[23-25].

CNS are fabricated directly on metal wires or strips. Metal wires of diameter 25 μm and lengths 4 to 5 cm are attached on a stainless steel platform and these wires act as the cathode during CNS growth. Fabrication takes place at a temperature of 650 °C and pressure of 6 Torr with the gas containing carbon and the etchant gas inserted in the chamber. The carbon-containing gas decomposes and deposits carbon on the metal wires. Since CNS is grown directly on the metal wires without the use of catalysts, the carbon grows uniformly on the substrate. This eliminates the fabrication step of requiring an insulating barrier during the fabrication process of CNS which is essential in VACNF and CNT fabrication processes[3, 22].

CNS has been fabricated on three different metal wires, silver, gold and titanium under growth times of 7.5 minutes, 5 minutes and 7.5 minutes. Stable electrodes can be obtained if growth times range between 3.5 to 7.5 minutes. Longer growth times may result in electrodes that are too noisy to use. The growth diameter was approximately 10 μm in each case, determined using SEM imaging. CNS wires have high aspect ratios which is an important characteristic for the detection of biological analytes. The carbon growth and the morphology of the CNS depend on the growth time and the substrate used. SEM images in Fig. 4 show that silver wire promotes more distinct spike-like

(a)   (b)   (c)

Fig. 4. CNS grown on different wires (a) gold (b) titanium (c) silver.

Table II. Catalysts and Substrates Used in the Fabrication of Carbon Nanostructures.

Catalyst	Substrate	Growth method	Carbon structure
Ni, Fe	Si, Glass	PECVD	CNT[24, 26]
Ni	Si, Quartz	PECVD	CNF[2, 27]
None	Ag, Au, Ti, Ta, Pd, Nb, Ni	PECVD	CNS[3]

carbon structures on the metal wire compared to gold and titanium. The features of CNS grown on gold and titanium have rounded structures compared to those grown on silver[30]. The summary of typical growth processes, types of sensor and substrates used for fabrication is shown in Table II.

6. Carbon Nanomaterial-Based Electrodes

6.1. *Structural Properties of the Electrodes*

Although carbon nanotubes have been widely used in biosensor research, carbon nanofibers represent a promising candidate for replacing carbon nanotubes. CNTs are made of concentric hollow graphene layers whereas CNFs are composed of graphene layers that form cups, plates or stacked cones. The internal nanofiber structure can be defined by the angle, α, between the fiber axis and the graphene layer. In CNTs, the value of α is zero as shown in Fig. 5[14]. Although there are significant differences in the structure of CNTs and CNFs, nanofibers are often named as nanotubes, however, the chemical and physical properties of nanotubes are different from those of nanofibers. CNFs have easier mass reproducibility, low production cost and are 100 times less expensive compared to CNTs[6]. CNTs have localized electrochemically active areas due to their extremely regular shapes, whereas the CNFs have active areas throughout its entire surface with many dangling bonds. This leads to improved sensitivity and responsiveness in CNFs compared to CNTs. CNFs are less cytotoxic compared to CNTs due to structural differences. CNFs also have improved hydrogen storage capability compared to CNTs,

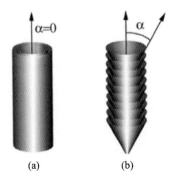

Fig. 5. Comparison between (a) CNT (b) CNF[13].

where hydrogen molecules are absorbed by Van der Waal's interaction to the surface atom of the carbon structures.

Although CNFs have many advantages over CNTs and both can be fabricated directly on silicon, integration with CMOS technology poses a number of challenges. CNS technology is one alternative that is used in order to gain batch reproducibility of the electrodes and the possibility of improved compatibility with the CMOS integration process[2, 3, 6]. CNSs demonstrate spike-like structure on the surface of the metal substrates. Research is still ongoing on the improvement of the properties of CNS. Strategies for nanostructure integration with CMOS integration are discussed in Sec. 6.2.

6.2. Integration with CMOS

Due to the many advantages, ranging from reduced power consumption, small volumes of biological fluids, and portability or implantability, integrated lab-on-a-chip systems are the aim of many biosensing applications research. The overall performance of the system will improve if the sensors and the readout electronics are in a compact integrated system. This is a direct result of lowering parasitics between the modules and incorporating signal processing directly into the sensing system. This results in faster measurement times and decreased size of the system. Although carbon nanostructures can be successfully fabricated directly on silicon substrates, the fabrication procedure of the readout circuit is not compatible with fabrication of nanostructures. The high thermal stress and the plasma arcing involved in the fabrication process of carbon nanostructures pose significant challenge to the integration procedures. The high temperature of approximately 700 °C creates an obstacle for the integration of CNFs and CNTs on a CMOS chip. The integration of nanostructures requires them to be encapsulated in an epoxy polymer and then released from the growth substrates in a separate fabrication step. The CNTs may then be aligned using micromanipulators However, in this process, the deterministic growth of CNTs due to catalytic growth process is lost. It is a time consuming, semi-manual method not suited to batch fabrication and requires realignment of the nanostructures thereby losing the advantage of the deterministic fabrication[28]. Employing dielectrophoretic processes is another way of integrating nanostructures with micro devices[29]. Using a two-step process where the CNF nanostructures are fabricated on a substrate in a PECVD chamber and then transferred to a new planar substrate. However, the transfer process of CNF onto another substrate requires handling of a tiny area, increasing the complexity, and resulting in poor chip area utilization and possible alignment issues. The method is illustrated for VACNFs in Fig. 6. CNS growth is uniform on planar electrodes, since a catalyst is not required in the growth process. In addition, the material is grown on metal wires which offer improved flexibility. The metal wire terminals can be connected to CMOS pads more efficiently using currently available bonding technologies[3, 29-30].

Reduction of the growth time and the temperature is extremely important for CMOS integration. The metal traces of the substrate can only withstand temperatures up to about ~525 °C for limited time without significant degradation in performance. The required

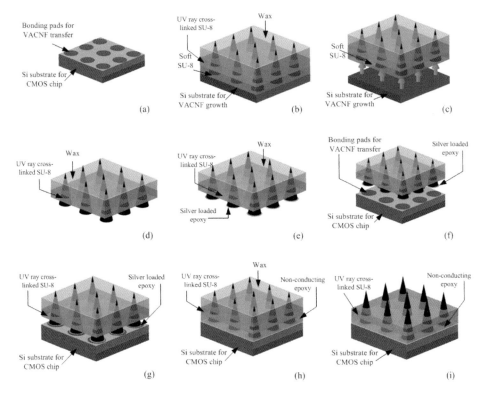

Fig. 6. Outline of VACNF transfer to CMOS chip process: (a) Cross section of the CMOS chip with bonding pads; (b) VACNF grown on Silicon wafer and coated with 2 layers of SU8 and 1 layer of wax. Upper SU8 is cross linked and lower one is soft SU8. Wax is used to increase mechanical strength during lift off; (c) VACNF along with 2 layers of Su8 and wax are lifted off from the silicon substrate; (d) Soft SU8 is etched away; (e) The base of each VACNF is painted with silver loaded epoxy; (f) The whole VACNF system is transferred to chip; (g) The VACNF is attached to the chip metal with silver loaded epoxy; (h) the hollow region between the cross linked SU8 and the over glass of the chip is filled with transparent non – conducting epoxy by flooding; (i) wax is washed out to expose more VACNF tip area.

growth temperatures have been reduced to as low as room temperature by using RF frequencies. An RF field can heat up the catalyst without increasing the temperature of the substrate which is a significant advantage over dc-PECVD process. The problem of plasma arcing in the PECVD process has been resolved by switching to CVD but this in turn increases the growth temperature[14]. However, if the growth time is reduced and moderate temperatures is used, the degradation of the CMOS circuits only needs to be characterized for a successful integrated fabrication process.

6.3. *Electrochemical Sensing on the Electrodes*

Electrochemical sensing is widely used because of its simple method to detect a variety of biological analytes. Neurotransmitters are chemical messengers that pass a signal between the cells and the neurons. Dopamine is a major test compound for neurochemical studies in neurotransmitter detections along with uric acid and ascorbic acid. Many enzyme reactions result in hydrogen peroxide (H_2O_2) as a byproduct. It acts as a reactive oxygen species in the body and is a byproduct of many industrial processes. Diabetes is a major cause of blindness, kidney failure, heart attack, and lower limb amputation and is typically characterized by glucose as an important analyte for detection. There are two methods to determine glucose: 1) nonenzymatic glucose detection 2) enzymatic electro-oxidation of glucose. Detection of alcohol has impact in the food and beverage industry and clinical analysis[7].

Different methods have been implemented for testing the detection limits of varying biological analytes. These include cyclic voltammetry, differential pulse voltammetry, fast scan cyclic voltammetry and amperometry. Cyclic voltammetry (CV) measurements show clear reduction and oxidation peaks where the potential is swept in a cyclic manner and the corresponding current is measured. In differential pulse voltammetry (DPV) the current is measured immediately before each potential change and the difference of the current is plotted with each potential. Fast scan cyclic voltammetry (FSCV) is similar to CV but uses a higher scan rate providing higher resolution. It is especially used in the detection of hormones, metabolites and neurotransmitters[7]. It can also be used to differentiate between multiple analytes measured at the same time.

All of the above mentioned methods are used to determine the detection limits of different carbon nanostructures[24-28]. Typical enzymes used in the electrochemical reactions and the nanostructures are summarized in Table III.

Table III. Carbon Nanostructure Electrodes Used on Different Analytes.

Sensor	Method	LOD	Analyte
CNT on diamond like carbon (DLC)[31]	CV	1.26±0.23 nM	Dopamine
CNF[32]	DPV	0.05 µM	Dopamine
CNS[3]	FSCV	8 nM	Dopamine
CNT/PB/CNT paste[33]	Amperometry	100 µM	H_2O_2
VACNF[34]	CV	66 µM	H_2O_2
CNT fiber/GO$_x$/CAD[35]	Amperometry	25 µM	Glucose
VACNF/GOx[21]	CV	40 µM	Glucose
GCE-MWCNT-Nf-HRP-SG/Chit-AOx-PEI[36]	Amperometry	5 µM	Ethanol
COOH-CNT/CFME[37]	FSCV	0.07 µM	Serotonin
CNF[32]	DPV	0.25 µM	Serotonin
CNF[38]	DPV	8.85×10^{-5} µM	Test DNA

7. Conclusion

Three types of carbon nanostructures suitable for lab-on-a-chip systems have been discussed. CNFs offer improved detection limits over CNTs, with CNS offering detection limits into the nM range. A range of analytes can be measured using the appropriate enzyme wiring technique. Integration with readout circuits is one of the biggest challenges facing the mainstream use of carbon nanostructures. Desirable features of the electrodes include their reproducibility and integration with external circuitry. Additionally, low substrate temperature growth is highly desirable. Integration with standard semiconductor integrated circuit technology requires reduced growth temperature and time. PECVD is a popular technology for growth of nanotubes, nanofibers and nanospikes due to the advantage of controlled lower temperature during the growth process. However, plasma arcing in the growth process can severely affect the integration of these electrodes with CMOS technologies. CNS grown on metal wires may be a solution to the integration problems.

Acknowledgment

A portion of this research, the PECVD growth of CNF, CNS and the SEM experiments (D.K.H.), was conducted at the Center for Nanophase Materials Sciences, ORNL, which is a DOE office of Science User Facility, under the user proposal number 2015-107.

References

1. J. Miyawaki, M.Yudasaka, T. Azami, Y. Kubo and S. Iijima, Toxicity of single-walled carbon nanohorns, *ACS Nano*, **2(2)**, pp. 213-226, 2008.
2. K. A. Mamun, F. S. Tulip, K. MacArthur, N. McFarlane, S. K. Islam, D. Hensley and I. Kravchenko, A robust VACNF platform for electrochemical biosensor, *IEEE Sensors 2013*, November 4-6, 2013.
3. A. G. Zestos, C. Yang, C. B. Jacobs, D. K. Hensley, and B. J. Venton, Carbon Nanospikes grown on Metal Wires as Microelectrode Sensors for Dopamine, *Analyst*, **140**, pp. 7283-7292, 2015. M. Pumera, A. Ambrosi, A. Bonanni, E. L. K. Chng and H. L. Poh, Graphene for electrochemical sensing and biosensing, *TrAC Trends in Analytical Chemistry*, **29**, issue 9, pp. 954-965, 2010.
4. M. Pumera, A. Ambrosi, A. Bonanni, E. L. K. Chng and H. L. Poh, Graphene for electrochemical sensing and biosensing, *TrAC Trends in Analytical Chemistry*, **29**, issue 9, pp. 954-965, 2010.
5. Y. Wang, Y. Li, L. Tang, J. Lu and J. Li, Application of graphene-modified electrode for selective detection of dopamine, *Electrochemistry Communications*, **11**, issue 4, pp. 889-892, 2009.
6. A. B. Islam, F. S. Tulip, S. K. Islam, T. Rahman and K. MacArthur, A mediator free bienzymatic glucose biosensor using vertically aligned carbon nanofibers (VACNF), *IEEE Sensors Journal*, **11**, no. 11, 2011.
7. C. Yeng, M. E. Denno, P. Pyakurel and B. J. Venton, Recent trends in carbon nanomaterial-based electrochemical sensors for biomolecules: A review, *Anal Chimica Acta*, **887**, pp. 17-37, Aug 7, 2015.
8. S. Singh, A. Singh, V. S. S. Bais, B. Prakash and N. Verma, Multi-scale carbon micro/nanofibers-based adsorbants for protein immobilization, *Material Science and Engineering* **C 38**, pp. 46-54, 2014.
9. M. Dervisevic, E. Cevik, Z. Durmus and M. Senel, Electrochemical sensing platforms based on the different carbon derivative incorporated interface, *Material Science and Engineering* **C 58**, pp. 790-798, 2016.

10. H. Ju, Z. Xueji and J. Wang, Nanobiosensing: Principles, Development and Application by *Springer LLC* 2011.
11. L. Matlock-Colangelo and A. J. Becumner, Recent progress in the design of nanofiber-based biosensing devices, *Lab Chip*, **12**, pp. 2612-2620, 2012.
12. R. Chen, Y. Li, k. Huo and P. K. Chu, Microelectrode arrays based on carbon nanomaterials: emerging electrochemical sensors for biological and environmental applications, *RSC Advances*, **3**, 18698, 2013.
13. Z. Wang and Z. Dai, "Carbon nanomaterial based electrochemical biosensors – an overview", *Nanoscale*, **7**, pp. 6420, 2015.
14. A. V Melechko, V. I. Merkulov, T. E. McKnight, M. A. Guillorn, K. L. Klein, D. H. Lowndes and M. L. Simpson, Vertically Aligned Carbon Nanofibers and Related Structures: Controlled Synthesis and Directed Assembly, *Journal of Applied Physics*, **97**, 041301, 2005.
15. Martin S. Bell, Kenneth B. K. Teo, Rodrigo G. Lacerda, W. I. Milne, David B. Hash and M. Meyyappan, Carbon nanotubes by plasma enhanced chemical vapor deposition, *Pure Appl. Chem.*, **28**, no. 6, pp. 1117-1125, 2006.
16. A. V. Melechko, K. L. Klein, J. D. Fowlkes, D. K. Hensley, I. A. Merkulov, T. E. McKnight, P. D. Rack, J. A. Horton and M. L. Simpson, Control of carbon nanostructure: From nanofiber toward nanotube and back, Journal of Applied Physics, **102**, 074314, 2007.
17. Y. Abdi, J. Koohsorkhi, J. Derakhshandeh, S. Mohajerzadeh, H. Hoseinzadegan, M. D. Robertson, J. C. Bennett, X. Wu, H. Radamson, PECVD grown carbon nanotubes on silicon substrates with a nickel-seeded tip-growth structure, *Material Science and Engineering* **C 26**, pp. 1219-1223, 2006.
18. Carbon nanotubes edited by Valentin N. Popov and Phillipe Lambin, Mathematics, Physics and Chemistry, **222**, Nato Science Series.
19. M. Endo, Y. A. Kim, M. Ezaka, K. Osada, T. Yanagisawa, T. Hayashi, M. Terrones and M. S. Dresselhaus, Selective and efficient impregnation of metal nanoparticles on cup-stacked-type carbon nanofibers, *Nano Letters*, **3**, no. 6, pp. 723-726, 2003.
20. E. Rand, A. Periyakaruppan, Z. Tanaka and D. A. Zhang, A carbon nanofiber based biosensor for simultaneous detection of dopamine and serotonin in the presence of ascorbic acid, *Biosensors and Bioelectronics*, **42C(1)**, pp. 434-438, November 2012.
21. K. A. Al Mamun, F.S. Tulip, K. McArthur, N. McFarlane and S.K. Islam, Vertically Aligned Carbon Nanofiber based Biosensor Platform for Glucose Sensor, *International Journal of High Speed Electronics*, **23**, Nos. 1 & 2, 1450006, 2014.
22. L. B. Sheridan, D. K. Hensley, N. V. Lavrik, S. C. Smith, V. Schwartz, C. Liang, Z. Wu, H. M. Meyer III, and a. J. Rondinone, Growth and Electrochemical Characterization of Carbon Nanospike thin film Electrodes, *J. Electrochem. Soc.*, **161**, issue 9, H558 – H563, 2014.
23. T. B. Ebbesen, H. Hiura, J. Fujita, Y. Ochiai, S. Matsui and K. Tanikagi, Patterns in the bulk growth of nanotubes, *Chemical Physics Letters*, **209**, no. 1 and 2, 1993.
24. K. Song, W. J. Wu, Y. S. Cho, G. S. Choi and D. Kim, The determining factors for the growth mode of carbon nanotubes in the chemical vapor deposition process, *Nanotechnology*, **15**, S592-S595, 2004.
25. V. I. Merkulov, A. V. Melechko, M. A. Guillorn, D. H. Lowndes and M. L. Simpson, "Growth rate of plasma-synthesized vertically aligned carbon nanofibers", *Chemical Physics Letters* **361**, pp. 492-498, 2002.
26. Z. F. Ren, Z. P. Huang, J. W. Xu, J. H. Wang, P. Bush, M. P. Siegal and P. N. Provencio, Synthesis of large arrays of well-aligned carbon nanotubes of glass, *Science*, **282**, issue- 5391, pp. 1105-1107, 1998.
27. Y. Yu, K. A. Al Mamun, A. S. Shanta, S. K. Islam and N. Mcfarlane, "Vertically Aligned Carbon Nanofibers as a Cell Impedance Sensor", IEEE Transactions on Nanotechnology, 2016.

28. B. L. Fletcher, T. E. McKnight, A. V. Melechko, D. K. Hensley, D. K. Thomas, M. N. Ericson and M. L. Simpson, Transfer of flexible arrays of vertically aligned nanofiber electrodes to temperature sensitive substrates, *Advanced Materials,* **18**, pp. 1689-1694, 2006.
29. M. S. Perez, B. Lerner, D. E. Resasco, P. D. P. Obregon, P. M. Julian, P. S. Mandolesi, F. A. Buffa, A. Boselli and A. Lamagna, Carbon nanotube integration with a CMOS process, *Sensors,* **10(4)**, pp. 3857-3867, 2010.
30. K. A. Al Mamun, J. Gu, D. K. Hensley, S. K. Islam, and N. McFarlane, Integration of Carbon Nanostructures on CMOS for Lab-on-a-chip Sensing, *IEEE International Symposium on Circuits and Systems CASFEST*, May 2016.
31. S. Sainioa, T. Palomäkia, S. Rhodeb, M. Kauppilac, O. Pitkänend, T. Selkäläd, G. Tothd, K. Kordasd, M. Moramb, J. Koskinene, T. Laurila, Carbon nanotube (CNT) forest on diamond-like carbon (DLC) electrode improves electrochemical sensitivity towards dopamine by two orders of magnitude, *Sens. Actuators*, B 211, pp. 177-186, 2015.
32. E. Rand, A. Periyakaruppan, Z. Tanaka and D. A. Zhang, A carbon nanofiber based biosensor for simultaneous detection of dopamine and serotonin in the presence of ascorbic acid, *Biosensors and Bioelectronics,* **42C(1)**, pp. 434-438, November 2012.
33. S. Husmann, E. Nossol, A. J. G. Zarbin, Carbon nanotube/Prussian blue paste electrodes: characterization and study of key parameters for application as sensors for determination of low concentration of hydrogen peroxide, *Sens. Actuators B Chem.*, **192**, pp. 782-790, 2014.
34. D. Suazo-Davila, J. Rivera-Melendez, J. Koehne, M. Meyyappan and C. R. Cabrera, Surface analysis and electrochemistry of a robust carbon-nanofiber based electrode platform, H_2O_2 sensor, *Applied Surface Science,* **384**, pp. 251-257, 2016.
35. Z. G. Zhu, L. Garcia-Gancedo, A. J. Flewitt, F. Moussy, Y. L. Li, W. I. Milne, Design of carbon nanotube fiber microelectrode for glucose biosensing, *J. Chem. Technol. Biotechnol.* **87**, pp. 256-262, 2012.
36. M. Das, P. Goswami, Direct electrochemistry of alcohol oxidase using multiwalled carbon nanotube as electroactive matrix for biosensor application, *Bioelectrochemistry,* **89**, pp. 19-25, 2013.
37. C. B. Jacobs, T. L. Vickrey, B. J. Venton, Functional groups modulate the sensitivity and electron transfer kinetics of neurochemicals at carbon nanotube modified microelectrodes, *Analyst,* **136**, pp. 3557-3565, 2011.
38. V. Vamvakaki, M. Fouskaki, N. Chaniotakis, Electrochemical biosensing systems based on carbon nanotubes and carbon nanofibers, *Anal. Lett.*, **40**, pp. 2271-2287.

Spatial Wavefunction Switched (SWS) FET SRAM Circuits and Simulation

Bander Saman[*], P. Gogna, El-Sayed Hasaneen[**], J. Chandy, E. Heller[***] and F. C. Jain

Department of Electrical and Computer Engineering,
University of Connecticut,
371 Fairfield Way, U-2157, Storrs, CT 06269-2157, USA
[*]Bander.Saman@uconn.edu

Current address: [*]Department of Electrical Engineering, Taif University, Taif, KSA
[**]Electrical Engineering Department, Aswan University, Egypt
hasaneen@engr.uconn.edu
[***]Synopsys Inc., Ossining, NY 10562, USA
evankheller@gmail.com

This paper presents the design and simulation of static random access memory (SRAM) using two channel spatial wavefunction switched field-effect transistor (SWS-FET), also known as a twin-drain metal oxide semiconductor field effect transistor (MOS-FET). In the SWS-FET, the channel between source and drain has two quantum well layers separated by a high band gap material between them. The gate voltage controls the charge carrier concentration in the quantum well layers and it causes the switching of charge carriers from one channel to other channel of the device. The standard SRAM circuit has six transistors (6T), two p-type MOS-FET and four n-type MOS-FET. By using the SWS-FET, the size and the number of transistors are reduced and all of transistors are n-channel SWS-FET. This paper proposes two different models of the SWS-FET SRAM circuits with three transistors (3T) and four transistors (4T) also addresses the stability of the proposed SWS-FET SRAM circuits by using the N-curve analysis. The proposed models are based on integration between Berkeley Short-channel IGFET Model (BSIM) and Analog Behavioral Model (ABM), the model is suitable to investigate the gates configuration and transient analysis at circuit level.

Keywords: SWS-FETs; multi-channel FETs; SRAM; VLSI.

1. Introduction

The SWS-FET allows the drain current flow in multiple channels in a single transistor. The device was first introduced by Jain *et al.* [1]. The two channels structure with two Si quantum well (W1 and W2) sandwiched between $Si_{0.5}Ge_{0.5}$ barriers is shown in Fig. 1.

The structure of two strained $Si/Si_{0.5}Ge_{0.5}$ layers in SWS-FETs offers carriers confinement based on applied gate voltage. This is due to band offsets and band edges alignment. The conduction band offset (ΔEc) in SWS-FET determine the degree of electron

[*]Corresponding author.

confinement in Si wells (lower well W2 and upper well W1). The conduction band offset is estimated as 0.15eV in strained Si/Si$_{0.5}$Ge$_{0.5}$ type II heterostructure [11].

Figure 2(a) shows the energy band diagram of the two quantum well SWS-FET type I heterostructure [2], the energy band diagram of type II heterostructure is shown in Fig. 2(b). From the Figs. 2(a) and 2(b), the conduction band offsets is equal to eΔEc = Ec Si-well-Ec SiGe-barrier = 1.04-0.89 = 0.15eV in strained Si/Si$_{0.5}$Ge$_{0.5}$ heterostructure.

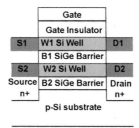

Fig. 1. The two channel SWS-FET structure.

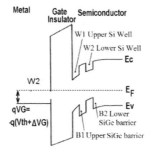

Fig. 2(a). The energy band diagram of Type I Heterostructure n-SWS-FET [2].

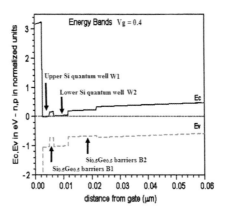

Fig. 2(b). The energy band diagram of Type II Heterostructure n-SWS-FET [2].

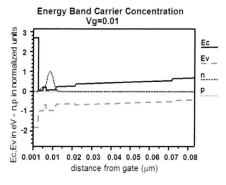
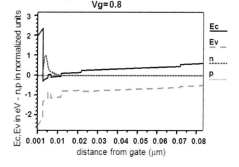

Fig. 3(a). Electrons in well W2 [3].   Fig. 3(b). Eectrons in well W1 [3].

Figure 3(a) and Fig. 3(b) present quantum simulations showing transfer of electron wavefunction from W2 to W1 as gate voltage is changed from 0.01V to 0.8V [3]. The fabrication process of SWS-FET can be done in same way as the conventional complementary metal-oxide-semiconductor (CMOS) FETs [2].

The switching modes are shown in Table 1. When the gate voltage (Vg) is applied in between zero and below threshold voltage of well 2 (V_{th2}) both wells W1 & W2 are in off mode, when Vg is set above V_{th2}, the electrons become confined in W2 which makes the current flows in W2 (I_{D2}). Once the gate voltage is increased and became greater than the threshold voltage of well 1 (V_{th1}), the electrons transfer from W2 to W1, and the current flows in W1 "I_{D1}" as well as I_{D2} drop off. And at time of Vg equals to the transition voltage (V_{UL}), the well-2 is in off mode [4].

Table 1. The operation mode.

Gate voltage Vg	Vg < vth2	vth2 < Vg < vth1	vth1 < Vg < VqL	Vg > VqL
Well 1	Off mode $I_{D1} \approx 0$	Off mode $I_{D1} \approx 0$	On mode $I_{D1} > 0$	On mode $I_{D1} > 0$
Well 2	Off mode $I_{D2} \approx 0$	On mode $I_{D2} > 0$	$\approx$ Off mode $I_{D2} \to 0$	Off mode $I_{D2} \approx 0$

2. SWS-FET Circuit Model

In terms of the device operation mechanism, a SWS-FET behaves as a MOS-FET in term of switching. The Eq. (1) illustrates the drain current for the MOS-FET, the same equation was developed for the SWS-FET to represent the drain current for well 1 and well 2 in Eq. (3) and Eq. (4) respectively [2-4].

$$I_{DS} = \left(\frac{W}{L}\right) C_{OX} \mu_n \left((V_{GS} - V_{TH})V_{DS} - \frac{V_{DS}^2}{2}\right) \tag{1}$$

$$I_{DS-well\,1} = \left(\frac{W}{L}\right) C_{ox}\mu_n \left((V_{G1S1} - V_{th1})V_{D1S1} - \frac{V_{D1S1}^2}{2}\right) \quad (2)$$

$$I_{DS-well\,2} = \left(\frac{W}{L}\right) C_{ox}\mu_n \left((V_{G2S2} - V_{th-well2})V_{D2S2} - \frac{V_{D2S2}^2}{2}\right) \quad (3)$$

The threshold voltage in W2 $V_{th\text{-well 2}}$ can be expressed as

$$V_{th-well2} = \begin{cases} V_{th2} & \text{When } V_{GSeff} < V_{UL} \\ V_{th2} + \alpha\,(V_{GSeff} - V_{UL}) & \text{When } V_{GSeff} > V_{UL} \end{cases}$$

where

$V_{th\text{-well2}}$ developed threshold voltage of W2
V_{th2} the threshold voltage of W2
V_{th1} threshold voltage of W1
V_{UL} the transition voltage
α the matching parameter, $\alpha = (V_{GS}-V_{UL})/(V_{th1}-V_{UL})$
V_{GS} the gate-source voltage
$V_{PolyEff}$ the voltage drop in the Poly Si gate
V_{GSeff} the effective gate-source voltage, $V_{GSeff} = V_{GS}-V_{PolyEff}$

Berkeley Short-channel IGFET Model (BSIM3) and Analog Behavioral Model (ABM) libraries are used to establish n-type SWS-FET model that can address the switching and changing in threshold voltages [5-7]. Figure 4 shows ABM block which represents the

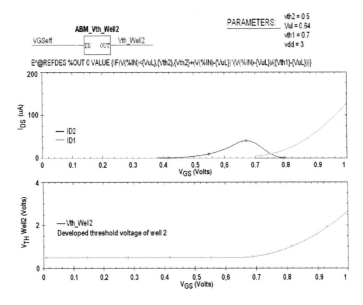

Fig. 4. ABM block for $V_{th\text{-well2}}$.

changing of $V_{th\text{-well 2}}$ and drains current switching, the input for the block is the effective gate-source voltage (V_{GSeff}) and the output is $V_{th\text{-well 2}}$. The changing of threshold offers the switching to BSIM transistors.

This model is set in hierarchical block and it ready to be used in Cadence-OrCAD CIS as shown in Fig. 5. In Fig. 6(a) the circuit-1 is configured to obtain the I_{DS}-V_{GS} characteristics of a two well single drain and single source SWS-FET, and the setting for this circuit is shown in Table 2. Figure 6(b) shows the I_{DS}-V_{GS} characteristic for three well single drain and single source SWS-FET with parameters of $V_{th1} = 2.25V$, $V_{th2} = 0.7V$, $V_{th3} = 0.2V$ and W/L = 10 μm / 5 μm. The simulation results show excellent performance on switching and output currents.

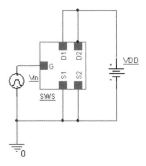

Fig. 5. Circuit 1 n-type SWS-FET hierarchical model.

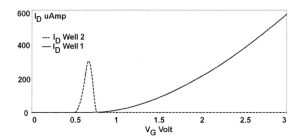

Fig. 6(a). Two channel SWS-FET I_{DS}-V_{GS} Characteristics.

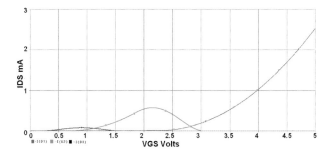

Fig. 6(b). Three channel SWS-FET I_{DS}-V_{GS} Characteristics.

Table 2. Circuit-1 model parameters.

Parameter	Value
L	5.0 μm
W	10 μm
Vth2	0.5 V
Vth1	0.7 V
VUL	0.64 V
VDD	3.0 V

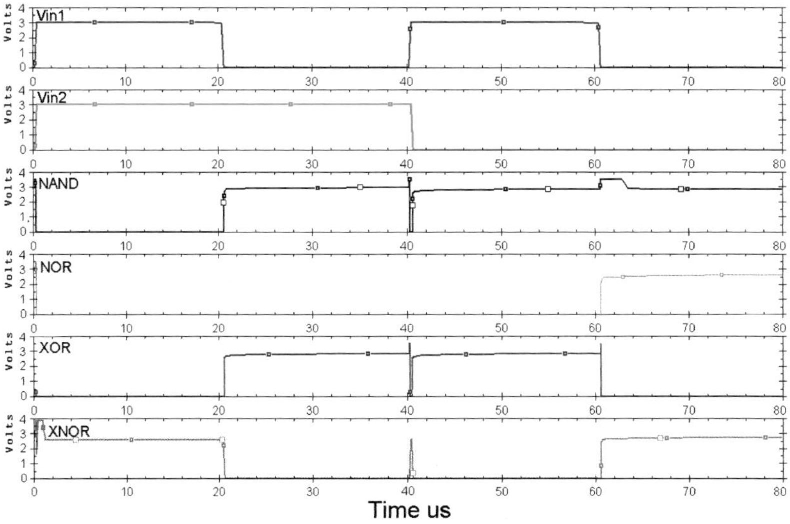

Fig. 7. n-channel SWS-FET gates.

Figure 7 shows the simulations of the n-channel SWS-FET logic gates NAND, NOR, XOR, and XNOR [7].

3. SRAM Circuit Model

A standard SRAM consists of n-type and p-type MOS-FET transistors for a memory implementation. The conventional 6T circuit has two inverter and two n-type MOS-FET access transistors as shown in Fig. 8(a) [8]. M1, M2, M3 and M4 transistors make a pair of inverters, connected in a loop. The other transistors M5 and M6 are used to control read and write. The circuit 8T in Fig. 8(b) has two more transistors to prevent the data from being disturbed during a read [8].

The proposed SWS-FET SRAM 4T cell is shown in Fig. 9(a); it has two conventional access transistors n-type MOS-FET (M1) and p-type MOS-FET (M2) for the write operation and one transistor n-type MOS-FET (M3) for the read operation. In this circuit SWS-FET (SWS1) works as memory cell and the source S1 and S2 are connected to the power supply (VDD) and ground (GND) respectively.

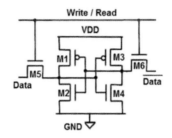

Fig. 8(a). CMOS SRAM 6T cell.

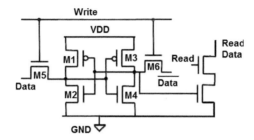

Fig. 8(b). CMOS SRAM 8T.

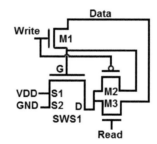

Fig. 9(a). SWS-FET SRAM 4T cell.

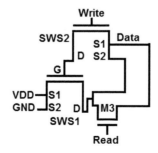

Fig. 9(b). SWS-FET SRAM 3T cell.

For the write operation, cell is accessed by turning ON M1 and turning OFF M2, in this case the data is passed to SWS1 gate (G) then to SWS1 drain (D). When the write signal became low M1 turns OFF and M2 turns ON, this conduction provides a loop between SWS1 drain, M2 and SWS1 gate which lets the data to store. For the read operation M3 turns ON and the stored data passes to the data line. Figure 9(b) is a modification of Fig. 9(a) to reduce the number of transistors. Here, the access transistors M1 and M2 are replaced by SWS2 to perform the switching of writing operation.

4. Simulation and Result

In this section, the performance of the proposed model is investigated and the SWS-FET has the parameters as in Table 3.

Table 3. Simulation model parameters.

Parameter	Value
L	5.0 μm
W	10 μm
Vth2	0.0 V
Vth1	0.5 V
VUL	0.1 V
VDD	3.0 V

The truth table of the SRAM is shown in Table 4. Figure 10 shows transient analysis of 6T COMS SRAM which matches with the truth table.

Table 4. Standard SRAM truth table.

Write signal	Data signal	Stored Data
0	0	Same State
0	1	Same State
1	0	0
1	1	1

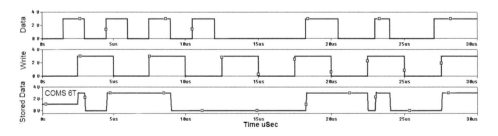

Fig. 10. The simulation CMOS T6 SRAM.

Figure 11 and Fig. 12 show the simulation for 4T and 3T SWS-FET SRAM respectively, and the model is based on SWS-BISM model [7]. The simulations match truth table and CMOS 6T simulation.

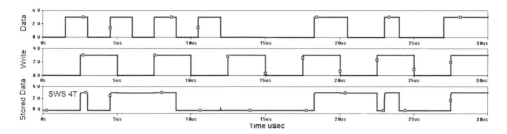

Fig. 11. The simulation of 4T SWS-FET SRAM.

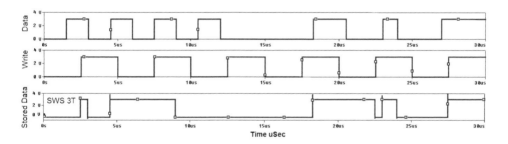

Fig. 12. The simulation of SWS-FET 3T SRAM.

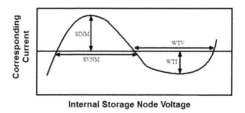

Fig. 13. The N-curve

The performance of the SRAM cell is normally measured by N-curve as shown in Fig. 13. Here, the test circuit has a voltage source in internal storage node to add a noise to the cell. The curve has all information about the read stability and the write ability.

The voltage between first and second zero crossing is called the static voltage noise margin (SVNM), and the peak current located between them is called as the static current noise margin (SINM). SVNM and SINM are defined as the maximum noise voltage and current at the input of the SRAM cell before its content changes. For better read stability SVNM & SINM should be larger [9-10].

In the N-curve, the voltage between the second and last zero crossing provides the write trip voltage (WTV) and the peak current located between them is called as the write trip current (WTI). They show the amount of voltage and current need to write in the cell, for better write ability and less operation power WTI has to be smaller [9-10].

Figure 14 shows the curves of standard 6T SRAM and both 4T/T3 SWS-FET SRAM. The standard 6T SRAM has better read stability and the SWS-FET SRAM has better write ability and less current in writing process. This is due to the reduction of number of the transistors. By increasing the threshold voltage of upper well (V_{th1}), the performance of SWS-FET SRAM improves in term of the read stability and the write ability as shown in Fig. 15 and Table 5.

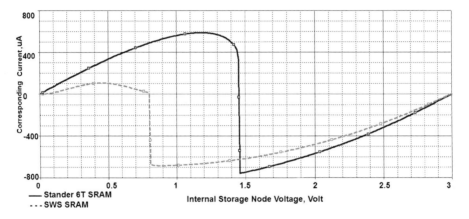

Fig. 14. The N-curve for standard and SWS-FET SRAM.

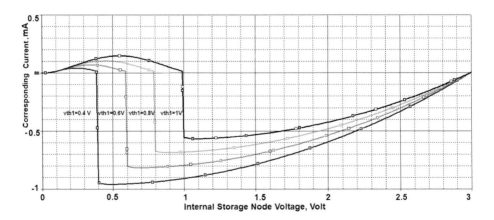

Fig. 15. The N-curve and the effect of changing of V_{th1}.

Table 5. the N-curve parameters of SVNM, SINM, WTV and WTI.

SRAM	SVNM V	SINM mA	WTV V	WTI mA
6T	1.47	0.58	1.53	0.75
SWS-FET SRAM, $V_{th1} = 0.4$ V	0.39	0.04	2.61	0.95
SWS-FET SRAM, $V_{th1} = 0.6$ V	0.58	0.07	2.42	0.81
SWS-FET SRAM, $V_{th1} = 0.8$ V	0.79	0.11	2.21	0.69
SWS-FET SRAM, $V_{th1} = 0.1$ V	0.98	0.15	2.02	0.57

5. Conclusion

In this paper, the simulations of I-V characteristics are presented for two and three quantum wells (QWs) n-channel SWS-FET. Also, the two wells n-channel SWS-FET is used to design two compact power efficient circuits of SRAM. The first proposed circuit 4T has one n-channel SWS-FET, two n-type MOS-FET and one p-type MOS-FET, the second circuit 3T has two n-channel SWS-FET and one n-type MOS-FET. The transient simulations are presented to verify the functionality of the proposed circuits T4/T3 and both circuits offer the write and the read operation as the standard 6T CMOS SRAM. The N-curve shows that the read stability and the write ability of both conventional 6T and proposed SRAM. The proposed cell has achieved improved write ability, and the read stability depends on the threshold voltage of upper well (W1).

References

1. F. Jain, J. Chandy, and E. Heller, Proc. "Lester Eastman Conf. on High Performance Devices", Int. J. High Speed Electronics and Systems, Vol. 20, pp. 641-652, September 2011.
2. P. Gogna, M. Lingalugari, J. Chandy, E. Heller, E-S. Hasaneen and F. Jain, "Quaternary Logic and Applications Using Multiple Quantum Well Based SWSFETs", International Journal of VLSI design & Communication Systems (VLSICS) Vol. 3, No. 5, October 2012.
3. F. Jain 2015 JEM DOI number
4. Jain, Miller, Suarez, Chan, Karmakar, Al-Amoody, Chandy, Heller, "Spatial Wavefunction Switched(SWS) InGaAs FETs with II-VI Gate Insulators", Journal of Electronic Materials, Vol. 40, No. 8, pp. 1717-1726, 2011.
5. BSIM3v3 Manual, (Final Version), web site: rely.eecs.berkeley.edu or 128.32.156.10.
6. Analog Behavioral Modeling Applications reference manual. Cadence – application note, December 2009.
7. B. Saman, P. Mirdha, M. Lingalugari, P. Gogna, and F. C. Jain., "LOGIC GATES DESIGN AND SIMULATION USING SPATIAL WAVEFUNCTION SWITCHED (SWS) FETS", International Journal for High-Speed Electronics and Systems, Vol. 24, Nos. 3 & 4 (2015) 1550008.
8. James Boley, Jiajing Wang and Benton H. Calhoun, "Analyzing Sub-Threshold Bitcell Topologies and the Effects of Assist Methods on SRAM Vmin", JLPEA 2012, 143-154.
9. Evelyn Grossar, Michele Stucchi, Karen Maex and Wim Dehaene, "Read stability and write-ability analysis of SRAM Cells for nanometer Technologies", IEEE Journal Of Solid-State Circuits, Vol. 41, No. 1, 2006.

10. Prabodh Kumar, Ashish Raman, "Analysis of Power and Stability of 7T SRAM Cell", International Journal of Advanced Research in Computer Science and Software Engineering, Volume 4, Issue 6, June 2014.
11. B. L. Stein *et al.*, "Band offsets in Si/Si$_{1-x-y}$Ge$_x$C$_y$ heterojunctions measured by admittance spectroscopy," Applied Physics Letters, Vol. 70, No. 25, pp. 3413-3415, Jun. 1997.

Carbon Nanotube Synthesis from Block Copolymer Deposited Catalyst

K. Woods, J. Silliman and T. C. Schwendemann[*]

Department of Physics,
Southern Connecticut State University, 501 Crescent St,
New Haven, CT 06515, USA
[*]schwendemant1@southernct.edu

A block copolymer/metal-salt solution was used to deposit metal nanoparticles on substrates, from which carbon nanotubes (CNTs) were grown in a chemical vapor deposition (CVD) chamber. Mono and hybrid catalysts of Fe, Ni, and Co-nitrates were tested, and separately Co, Ni, and Cu-chlorides. In both cases cobalt/cobalt-hybrids produced the highest density of multi-wall carbon nanotubes (MWCNTs). Slight vertical growth, though sparse, was observed after growth at 800°C from a nickel catalyst on single-crystal aluminium oxide (~130nm diameter).

Keywords: carbon nanotube; block copolymer.

1. Introduction

The goal of this research was to find a method of nano-particle synthesis and the appropriate CVD conditions for the growth of an aligned array of single wall carbon nanotubes (SWCNTs). CNTs are seamless one-atom thick cylinders of covalently bonded carbon atoms[1], and can grow tightly nested within one another in which case they are termed multi-walled CNTs (MWCNTs). CNTs are already used commercially to increase the strength and/or conductivity of materials as a composite[2], and to increase power to weight ratios for lithium-ion batteries. Better control over CNT diameter, distribution, and alignment on surfaces would allow for many new or more efficient devices. Vertically aligned CNT's will likely be utilized in future fuel cells[3], solar cells, and sensors/microchips. CNTs can be grown in a chemical vapor deposition (CVD) chamber using metallic nanoparticles as a catalyst[4,5]. Best alignment of CNTs is often achieved using plasma enhanced CVD; our aim was to obtain alignment using a standard CVD chamber.

There are many ways of depositing metallic nanoparticles on a surface, with varying materials and degrees of control. One method to arrange nanoparticles on a surface is to create a solution of metal-salts and block copolymers (BCP). The metal-salts bond to the BCP chains, then the BCPs self-organize into an array of ordered dots on a flat surface as

[*]Corresponding author.

the solvent evaporates. Metallic-nanoparticles form in place of the dots after the organic matter has been burned off with oxygen plasma[6,7,8].

2. Method

First, 0.1g polystyrene-b-poly(4-vinyl pyridine) (PS-b-P4VP; 21,000-b-18,000) is added to 25mL of toluene and stirred at 70°C for 6 hours. Then a transition metal–salt is added in a 0.25 molar ratio of metal ions to pyridine groups (of the block copolymer) and stirred at 30°C for 18 hours. Mono and hybrid catalysts of iron (Fe), nickel (Ni), and cobalt (Co) nitrates were tested, and separately Co, Ni, and copper (Cu) chlorides.

Using glass syringes, the solutions are spin-cast on a substrate at 1250 RPM for 1 minute (prior to this the substrates sat in acetone for 5 minutes and were rinsed in methanol). Thermally oxidized silicon (Si) was used with both nitrates and chlorides, polycrystalline aluminum oxide (pc Al_2O_3) for the nitrates, and single crystal aluminum oxide (sc Al_2O_3) with the chlorides. Alternatively, a drop may be deposited on a substrate and then sandwiched with another substrate and slid apart to deposit a thin film of solution on each.

The samples are then exposed to oxygen plasma for 30 minutes to remove the block copolymer. Lastly, the samples are placed in a CVD chamber and a gas (95% Ar 5% H_2) is flown through the chamber to displace oxygen as the samples are heated to a set temperature. A carbon-containing gas (ethylene) is then added to the flow to cause CNT growth. After a set amount of time the carbon source is turned off and the chamber is allowed to cool down.

The following CVD parameters were tested: temperature (°C: 600, 700, 750, 800), total gas flow (Std. Liters per Minute (SLM); 0.66, 1, 1.2, 1.5, 2), time (min: 5, 10, 30, 45), oxygen displacer (Argon/Hydrogen, Ethylene), and sample orientation in CVD (slanted, recessed, or bridged and open or partially covered).

3. Results and Discussions

3.1. *Fe, Ni, Co – Nitrates*

As evident from the scanning electron microscope (SEM) images, carbon nanotube growth was seen from many experimental conditions. Cobalt grew medium-density forests of nanotubes under all conditions tested (Fig. 1). At 750°C for 45 minutes and 1 std. liter/min (SLM) combined, Co, Fe/Co, and Fe/Ni all effectively catalyzed the growth of MWCNTs. Under the same conditions with the sample boat partially covered, growth was significantly hindered for Fe/Co while Co was unaffected. Sample orientation had no noticeable effect on CNT alignment or density. The CNT's are multi-walled CNT's and had average diameters of about 30 nm (Co, Fe/Co) and 35 nm (Fe/Ni), with little variation between CVD batches.

Figure 2 shows block copolymer dots with non-uniform sizes, preventing large-scale ordering. This did not change much after oxygen plasma treatment and very little agglomeration was observed pre-CVD. The BCP dots were larger and closer together than expected, leading to larger nanoparticles and thus larger diameter (and multi-wall) CNTs. Though Co, Fe/Co, and Fe/Ni performed well as catalysts, a significant fraction of nanoparticles failed to produce CNTs. It is likely that the nanoparticles were too close to each other and may have agglomerated under CVD conditions; bare patches of substrate could sometimes be seen with isolated CNTs. It is difficult to be certain, but the minority of CNTs appear to have bright tips (indicating metallic nanoparticles), suggesting that the root-growth mechanism was dominant.

After 5 minutes at 750°C and 1.2 SLM combined gas-flow an Fe/Ni catalyst was already growing nanotubes, with Co requiring up to 5 more minutes to begin growth. Fe/Ni hybrid catalysts performed well (Fig. 3), but alone neither Fe nor Ni led to good nanoparticle formation and neither was able to grow more than isolated CNTs.

Instead of spin-casting, an alternative preparation may sandwich a drop of BCP-metal-salt solution between two substrates and slide them apart at a constant rate. In tests by hand, inconsistent sliding produced stripes of good and poor nanoparticle formation. For either method, filtering the BCP solution through sub–micron pores would prevent deposition of undissolved solids.

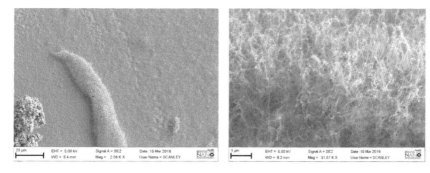

Fig. 1. Carbon nanotubes grown by Cobalt deposited on substrates at 750°C and 1.5 SLM. The images on the left shows area of dense growth and sparse growth. The image on the right shows a high resolution image of the grown nanotubes.

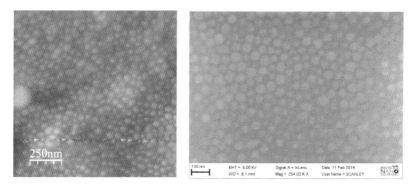

Fig. 2. The figure on the left shows an AFM image of the block copolymers with nanoparticles. And SEM the image on the right shows the nanoparticles after oxygen plasma to remove the block copolymer.

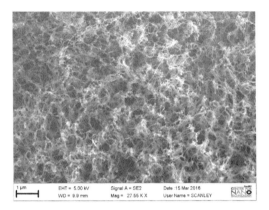

Fig. 3. MWCNT grown from Fe/Ni nanoparticles at 750°C.

3.2. Co, Ni, Cu – Chlorides

A nickel catalyst grew the largest diameter MWCNTs observed (~200nm grown at 800°C and 1 SLM combined on silicon), though growth was very sparse. Cobalt grew fairly dense forests of large diameter (130nm) MWCNTs on aluminum oxide at 800°C. Growth from cobalt was less likely on silicon, and the diameters of the MWCNTs were much larger (3×) when grown at 800°C compared to 750°C for both silicon and single crystal aluminum oxide; very similar trends were observed for nickel and slight vertical growth, though sparse, was observed on single-crystal aluminium after growth at 800°C and 1 SLM combined (~125nm diameter, Fig. 4). The CNTs grown at 800°C also had a slightly corrugated appearance in the SEM; when imaged in a transmission electron microscope (TEM) a very textured appearance could often be seen as well as thin lines along the center of the MWCNTs. This may indicate that some of the nanotubes have some degree of graphenization and/or hollow cores of several nanometers. Graphenated CNTs have been used as compressible aerogels in super-capacitors[9], and may one day

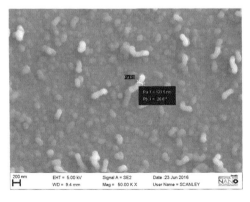

Fig. 4. Nickel catalyzed MWCNT growth on aluminum oxide grown at 800°C and 1 SLM, showing very sparse growth under these conditions.

form the key component in ultra-capacitors as well. Ultra-capacitors would charge up very fast like a regular capacitor, but release lots of energy slowly like a battery.

In general, growth was slightly denser for a total gas-flow of 2 SLM ethylene compared to 1 SLM ethylene. Alone, Cu and Ni were poor catalysts, but a hybrid catalyst composed of them performed significantly better avg. diameter ~30nm. If Co were substituted for either element catalyst activity improved further, with CNT diameter slightly increasing to almost 40nm for Co/Cu hybrid catalyst.

4. Conclusion

Well formed BCP dots likely led to well-formed nanoparticles, which may be more likely to produce CNTs than poorly formed nano-particles; proper thickness of the deposited BCP solution is important for good dot formation. A controlled method of sandwiching solution between substrates and sliding them apart could be used for samples too large to spin-cast.

For the metal-nitrates, cobalt performed the best, and if the catalyst contained cobalt the plasma treatment step was not required for the synthesis of CNTs (although growth was slightly hindered if the plasma was skipped). Cobalt performed comparably across 600 to 800°C with combined SLM ranging from 1 to 2. Fe/Co and Fe/Ni performed relatively well in general, Ni/Co did poorly, and individually both Fe and Ni did very poorly. Alignment was not achieved and these catalysts were only tested on silicon and polished poly-crystalline aluminum oxide, with all catalysts performing significantly worse on the aluminum oxide. Future work should test these catalysts on single crystal aluminum oxide including higher temperatures and a type of block copolymer which gives more space between dots.

For the metal-chlorides, again Co (and Co-hybrids) performed the best out of the conditions tested. At 800°C for 30 minutes with 1 SLM combined and Ar/H_2 pre-flow both Co and Ni on aluminum oxide performed significantly better than at 750°C. Growth for Co on aluminum oxide at 750°C was improved by using 1 SLM total of ethylene, and slightly further with 2 SLM total of ethylene. Mono-catalytic substrates of Co and Ni produced MWCNTs at 800°C with three to four times the diameter compared to those grown at 750°C on both silicon and single crystal aluminum oxide.

Of all scenarios tested, a Ni-chloride catalyst on aluminum oxide grown at 800°C for 30 minutes with 1 SLM combined gas-flow showed the best signs of vertical growth, though coverage was sparse. It is possible that the large diameter (~125nm) of the nanotubes along with substrate interaction encouraged vertical growth. Pure copper was unable to grow CNTs, but if combined with Ni the catalytic activity was higher than that for pure Ni. Future work should explore the relationship between CNT diameter and growth temperature (750°C and up) for Co and Ni as mono and hybrid catalysts with Cu and/or Fe with the aim of producing large diameter vertically aligned MWCNTs. Single-crystal aluminium oxide was proven superior to thermally oxidized silicon for metal-chloride catalysis of MWCNTs, and the addition of other catalytic elements may alter substrate interaction and further enhance vertical alignment, possibly through encouraging tip-growth of the CNTs.

Acknowledgment

The authors acknowledge support from the National Science Foundation [NSF] funded Materials Research Science and Engineering Center [MRSEC] Center for Research on Interface Structures and Phenomena [CRISP] via support and use of CRISP facilities (CRISP NSF DMR 1119826. Additional support is acknowledged from the SCSU Department of Physics, the CSCU-CNT Center for Nanotechnology, and NASA CT Space Grant Pro-Sum #962.

References

1. S. Iijima, *Nature* **354**, 56 (1991).
2. Z. F. Ren, Z. P. Huang, J. W. Xu, J. H. Wang, P. Bush, M. P. Seigal, P. N. Provencio, *Science* **282**, 1105 (1998).
3. W. Li, Xin Wang, Z. Chen, M. Waje, and Yushan, *Langmuir* **21**, 9386 (2005).
4. O. A. Louchev, Y. Sato, and H. Kanda, *Applied Physics Letters* **80**, 2752 (2002).
5. C. Jenkins, M. Cruz, J. Depalma, M. Conroy, B. Benardo, M. Horbachuck, T. Sadowski, C. Broadbridge, T. C. Schwendemann, *International Journal of High Speed Electronics and Systems* **23**, no 01n02 (2014).
6. J. Q. Lu, *J. Phys. Chem. C* **112**, 10344 (2008).
   S. Bhaviripudi, A. Reina, J. Qi, J. Kong, and A.M. Belcher, *Nanotechnology* **17**, 5080 (2006).
7. R. D. Bennet, A. J. Hart, and R. E. Cohen, *Adv. Mater.* **18**, 2274 (2006).
8. E. Wilson, and M. F. Islam, *Applied Materials and Interfaces* **7**, 5612 (2015).

Dynamical X-Ray Diffraction Analysis of Triple-Junction Solar Cells on Germanium (001) Substrates

F. A. Althowibi
Electrical Engineering Department,
Taif University, Taif, Saudi Arabia
fhdamer@gmail.com

J. E. Ayers
Electrical and Computer Engineering Department,
University of Connecticut,
Storrs, Connecticut 06269-4157, USA
john.ayers@uconn.edu

We demonstrate the dynamical x-ray diffraction analysis of metamorphic triple-junction solar cells grown on Ge (001) substrates. The solar cells investigated involve an $In_{0.67}Ga_{0.33}P$ top cell, an $In_{0.17}Ga_{0.83}As$ middle cell, and a Ge bottom cell. A graded buffer layer is inserted between the bottom and middle cells for the purpose of accommodating the lattice mismatch. Linearly-graded, step-graded, and S-graded compositional profiles were considered for this buffer layer. The x-ray rocking curve analysis for a number of *hkl* reflections including 004, 113, 116, 044, 026, and 117 was conducted for the case of Cu Kα_1 radiation. We show that the use of non-destructive x-ray analysis allows determination of the threading dislocation densities in the top two cells. In the cases of S-graded or step-graded buffer layers, the buffer threading dislocation density could also be estimated.

Keywords: x-ray diffraction; dynamical diffraction; metamorphic heterostructures; dislocations.

1. Introduction

Multi-junction solar cells are of great interest for space applications as well as terrestrial concentrator installations, and efficiencies exceeding 40% have been demonstrated. Metamorphic realizations are preferred, because this approach allows great freedom to optimize the band-gaps of the individual cells, but metamorphic growth results in threading defects which have a deleterious effect on efficiency.

Traditional methods of x-ray rocking curve analysis involve comparison of experimental profiles to dynamical [1-6] and kinematical [7-11] simulations and these allow depth profiling of strain and compositions in pseudomorphic structures free from dislocations. This approach has recently been extended to metamorphic structures using the phase-invariant [12-15] and mosaic crystal [16] models.

In the present work, we have utilized the mosaic crystal model to study the analysis of the threading dislocation densities in metamorphic triple-junction solar cells grown on Ge (001) substrates with the inclusion of linearly-graded, step-graded, and S-graded [17-19] buffer layers. In the S-graded buffer layers, the mean parameter was assumed to be the midpoint in the buffer layer thickness, and the standard deviation parameter was assumed to be 0.22µm. We consider several hkl x-ray diffraction profiles measured using Cu kα_1 radiation.

2. Theory

For a perfect and infinitely thick semiconductor crystal, the diffraction profile may be calculated by the solution of the Takagi-Taupin equation [2-3] for dynamical diffraction,

$$-i\frac{dX}{dT} = X^2 - 2\eta X + 1 \tag{1}$$

in which X is the complex scattering amplitude, η is the deviation parameter, and T is the thickness parameter calculated by

$$T = h\frac{\pi\Gamma\sqrt{F_{HS}\,F_{\bar{H}S}}}{\lambda\sqrt{|\gamma_0\gamma_H|}} \tag{2}$$

where h is the depth measured from the diffracting surface, and F_{HS} and $F_{\bar{H}S}$ are the substrate structure factors for the hkl and $\bar{h}\bar{k}\bar{l}$ reflections and $\Gamma = r_e\lambda^2/(\pi V)$, where r_e is the classical electron radius, $2.818\times 10^{-5}\text{Å}$, λ is the x-ray wavelength, V is the unit cell volume, and γ_0 and γ_H are the direction cosines for the incident and reflected beams with respect to the inward surface normal. The resulting scattering amplitude for the substrate is described by the Darwin-Prins formula [20] as

$$X_0 = \eta_s - Sign(\eta_s)\sqrt{\eta_s^2 - 1} \tag{3}$$

where the deviation parameter for the substrate is given by

$$\eta_s = \frac{-(\gamma_0/\gamma_H)(\theta - \theta_{BS})sin(2\theta_{BS}) - 0.5(1 - \gamma_0/\gamma_H)\Gamma F_{0S}}{\sqrt{|\gamma_0/\gamma_H|}\,C\Gamma\,\sqrt{F_{HS}\,F_{\bar{H}S}}} \tag{4}$$

and θ_{BS} is the Bragg angle for the substrate, θ is the actual angle of incidence on the diffracting planes, F_{0S} is the substrate structure factor for 000 scattering, and C is the polarization factor.

In a metamorphic device structure, the diffuse scattering associated with dislocations must be considered. For practical device structures in which the dislocations take on irregular configurations and must be treated using their ensemble averages, the phase invariant and mosaic crystal models may be used to calculate the x-ray diffraction profiles. Here we apply the more generally applicable mosaic crystal model, which is summarized briefly below. The reader is referred to Ref. [16] for additional details.

For a defect-free semiconductor heterostructure divided into a number of sublayers or lamina, solution of the Takagi-Taupin equation [2-3] gives the scattering amplitude at each individual sublayer. For the nth layer in the stack, the scattering amplitude at the top of the layer, X_n, is related to the scattering amplitude at the bottom of the layer, X_{n-1}, by

$$X_n = \eta_n + \sqrt{\eta_n^2 - 1} \frac{(S_{1n} + S_{2n})}{(S_{1n} - S_{2n})} \tag{5}$$

where

$$S_{1n} = \left(X_{n-1} - \eta_n + \sqrt{\eta_n^2 - 1}\right) exp\left(-iT_n\sqrt{\eta_n^2 - 1}\right) \tag{6}$$

and

$$S_{2n} = \left(X_{n-1} - \eta_n + \sqrt{\eta_n^2 - 1}\right) exp\left(iT_n\sqrt{\eta_n^2 - 1}\right). \tag{7}$$

η_n is the deviation parameter and the thickness parameter is

$$T_n = h_n \frac{\pi \Gamma \sqrt{F_{Hn} F_{\bar{H}n}}}{\lambda \sqrt{|\gamma_0 \gamma_H|}} \tag{8}$$

where h_n is the thickness, and F_{0n}, F_{Hn} and $F_{\bar{H}n}$ are the 000, hkl and $\overline{hkl}$ structure factors for the nth sublayer.

For a defected semiconductor device structure, the crystal is distorted by the presence of dislocations which introduce variations of orientations [21,23-24] and interplanar spacings [22-24]. To include these variations, the structure is treated as a mosaic of $N_\alpha \times N_\beta$ crystallites in α and β space, where α represents the angle-scale distribution associated with the orientational mosaicity and β represents the angle-scale distribution associated with the mosaicity of the interplanar spacing. At each individual crystallite, the incidence angle is adjusted for angular variations through α and the Bragg angle is modified for strain variations through β. The scattering amplitude is iteratively computed for each of these crystallites at each angle of interest. For the nth lamina of the ijth crystallite, the deviation parameter η_{nij} is calculated by

$$\eta_{nij} = \frac{-(\gamma_0/\gamma_H)(\theta - \theta_{Bn} + \alpha_i - \beta_j)sin(2\theta_{Bn}) - 0.5(1 - \gamma_0/\gamma_H)\Gamma F_{0n}}{\sqrt{|\gamma_0/\gamma_H|}\, C\Gamma\, \sqrt{F_{Hn} F_{\bar{H}n}}} \tag{9}$$

while the scattering amplitudes are calculated iteratively by

$$X_{nij} = \eta_{nij} + \sqrt{\eta_{nij}^2 - 1}\, \frac{(S_{1nij} + S_{2nij})}{(S_{1nij} - S_{2nij})} \tag{10}$$

where

$$S_{1nij} = \left(X_{nij-1} - \eta_{nij} + \sqrt{\eta_{nij}^2 - 1}\right) exp\left(-iT_n\sqrt{\eta_{nij}^2 - 1}\right) \quad (11)$$

and

$$S_{2nij} = \left(X_{nij-1} - \eta_{nij} + \sqrt{\eta_{nij}^2 - 1}\right) exp\left(iT_n\sqrt{\eta_{nij}^2 - 1}\right) \quad (12)$$

and the thickness parameter is

$$T_n = h_n \frac{\pi\Gamma\sqrt{F_{Hn}\,F_{\bar{H}n}}}{\lambda\sqrt{|\gamma_0\gamma_H|}} \quad (13)$$

where h_n is the thickness, and F_{0n}, F_{Hn} and $F_{\bar{H}n}$ are the 000, hkl and $\bar{h}\bar{k}\bar{l}$ structure factors for the nth sublayer. The x-ray diffraction profile is calculated by adding weighted intensity contributions from the $N_\alpha \times N_\beta$ crystallites in α and β space; therefore

$$I = \sum_i \sum_j |X_{nij}|^2 \cdot W_{\alpha i} \cdot W_{\beta j} \quad (14)$$

where $W_{\alpha i}$ and $W_{\beta j}$ are the weighting functions accounting for angular and interplanar spacing variations, and their distributions are assumed to be Gaussian given by [21-24]

$$W_{\alpha i} = exp(-\alpha_i^2/2\sigma_\alpha^2) \quad (15)$$

and

$$W_{\beta j} = exp(-\beta_j^2/2\sigma_\varepsilon^2) \quad (16)$$

where $\alpha_i = -N_\sigma\sigma_\alpha + iN_\alpha\sigma_\alpha/2N_\sigma$ and $\beta_j = -N_\sigma\sigma_\beta + jN_\beta\sigma_\beta/2N_\sigma$, where N_σ is the number of standard deviations used in the two distributions and i and j are integers. The standard deviations are [22-24] given by

$$\sigma_\alpha = b\sqrt{\pi D/2} \quad (17)$$

and

$$\sigma_\varepsilon = 0.127b \cdot \sqrt{D|ln(2\times10^{-7}cm\sqrt{D})|} \cdot tan(\theta_B) \quad (18)$$

where D is the dislocation density, and b is the Burgers vector for the dislocations.

This approach serves as the basis for the x-ray characterization of metamorphic structures, allowing determination of the depth profiles of strain, composition, and dislocation density, and in this work we have applied it to triple-junction metamorphic solar cells grown on Ge (001) substrates.

3. Results and Discussion

We have calculated the dynamical rocking curves from metamorphic triple-junction solar cells grown on Ge (001) substrates for a number of *hkl* reflections for the case of Cu Kα_1 radiation (λ = 0.1540594 nm). Figure 1 shows the nominal design of the metamorphic triple-junction solar cells considered with an $In_{0.65}Ga_{0.35}P$ top cell, an $In_{0.17}Ga_{0.83}As$ middle cell, and a Ge bottom cell. In the work of Guter *et al.* [25] the top two cells were designed to be lattice-matched to one another, but do not match the Ge substrate, necessitating the use of a graded buffer layer in between the bottom and middle cells. In their experimental work Guter *et al.* [25] used a step-graded buffer, but here we also calculated rocking curves for structures having linearly-graded and S-graded buffer layers. Thirty sublayers were used to model the linearly-graded buffer; determination of the required number of sublayers is discussed further in [16]. Because the top and middle cells in the nominal design had compositions of $In_{0.65}Ga_{0.35}P$ and $In_{0.17}Ga_{0.83}As$, respectively, to obtain approximate lattice matching, it will be difficult to distinguish the rocking curve peaks for these layers. Here we have modified the composition in the top cell to $In_{0.67}Ga_{0.33}P$ so that the rocking curves from the top two cells may be resolved, and this enables the characterization of the individual threading dislocation densities in the top two cells. In the nominal design, these peaks would not be resolved, and it would only be possible to determine the weighted average threading dislocation for the top two cells as a whole. However, this less detailed characterization still provides valuable feedback to the crystal grower, and the average threading dislocation density so obtained would be valuable in evaluating different graded buffer approaches.

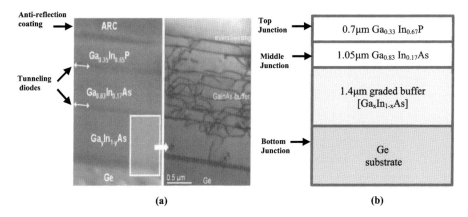

Fig. 1. $In_yGa_{1-y}P/In_xGa_{1-x}As$/Ge triple junction solar cell containing a graded $In_xGa_{1-x}As$ buffer layer. (a) Cross-sectional TEM images of triple-junction solar cell considered in experimental work by Guter *et al.* [25]. Reprinted from W. Guter, J. Schone, S. P. Philipps, M. Steiner, G. Siefer, A. Wekkeli, E. Welser, E. Oliva, A. W. Bett, and F. Dimroth, J. Appl. Phys. Lett., 94, 223504 (2009), with the permission of AIP Publishing. (b) Schematic of the triple-junction solar cell heteroepitaxial structure used for x-ray analysis. The composition of the top layer has been adjusted to 67% indium so that the peaks from the top two cells may be resolved.

All calculated rocking curves were smoothed by convolution with a Gaussian instrumental function having a FWHM of 12 arc seconds, which represents the divergence for a two-crystal, four-reflection Ge (220) Bartels monochromator [26-27].

First, we calculated the baseline 004 diffraction profiles for dislocation-free triple-junction solar cells containing the three types of buffer layers. Figures 2-4 show the 004 rocking curves for solar cells with step-graded, linearly-graded and S-graded buffer layers, respectively, calculated with the assumption of zero dislocation density. Three distinct peaks corresponding to the three solar cells are evident. The step-graded and S-graded buffer layers exhibit additional resolved peaks, whereas the diffracted intensity from the linearly-graded buffer appears as a single wide band. Five peaks are resolved for the step-graded layer, corresponding to the five steps, while two peaks are observed from the S-graded buffer, corresponding to the top and bottom compositions.

Next we considered the effect of dislocations on the diffraction profiles. Figures 5-7 illustrate the 004 rocking curves for step-graded, linearly-graded, and S-graded triple-junction solar cells, in which the threading dislocation density was assumed to be 0, 10^5, 10^6, 10^7, 10^8, and 10^9 cm^{-2}. For the metamorphic uniform-composition layers utilized in the middle and top cells, the peak widths increase monotonically with the dislocation density, so correlation of the two quantities should allow estimation of the defect density from the measured width. The same is true for the individual layers in the step-graded buffer, to the extent that the peaks do not overlap significantly because of the broadening.

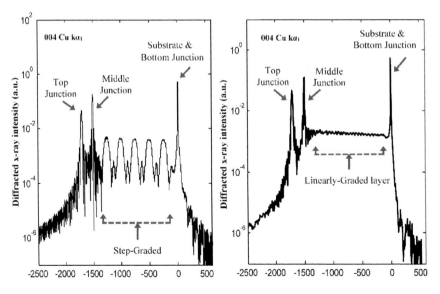

Fig. 2. The 004 rocking curves for a $Ga_{0.33}In_{0.67}P/In_xGa_{1-x}As/Ge$ solar cell with a 1.4μm step-graded buffer (7 steps, 200 nm each), assumed to be dislocation free.

Fig. 3. The 004 rocking curves for $Ga_{0.33}In_{0.67}P/In_xGa_{1-x}As/Ge$ solar cell with a 1.4μm linearly-graded buffer, assumed to be dislocation free.

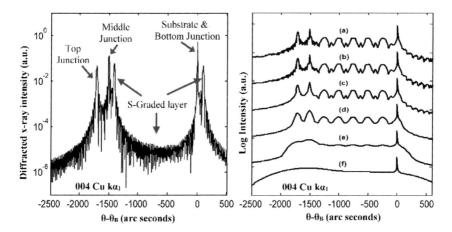

Fig. 4. The 004 rocking curves for $Ga_{0.33}In_{0.67}P/In_{0.17}Ga_{0.83}As/In_xGa_{1-x}As/Ge$ solar cell with a 1.4μm S-graded buffer layer, assumed to be dislocation free.

Fig. 5. The 004 rocking curves for a $Ga_{0.33}In_{0.67}P/In_xGa_{1-x}/Ge$ solar cell with a 1.4μm step-graded buffer layer (7 steps, 200 nm each) for the following dislocation densities: (a) $D=0$ cm^{-2}. (b) $D=10^5$ cm^{-2}. (c) $D=10^6$ cm^{-2}. (d) $D=10^7$ cm^{-2}. (e) $D=10^8$ cm^{-2}. (f) $D=10^9$ cm^{-2}.

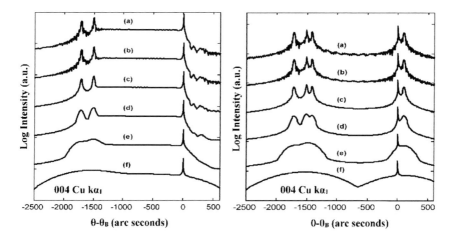

Fig. 6. The 004 rocking curves for a $Ga_{0.33}In_{0.67}P/In_xGa_{1-x}As/Ge$ solar cell with a 1.4μm linearly-graded buffer, for the following values of dislocation density: (a) $D=0$ cm^{-2}. (b) $D=10^5$ cm^{-2}. (c) $D=10^6$ cm^{-2}. (d) $D=10^7$ cm^{-2}. (e) $D=10^8$ cm^{-2}. (f) $D=10^9$ cm^{-2}.

Fig. 7. The 004 rocking curves for $Ga_{0.33}In_{0.67}P/In_xGa_{1-x}As/Ge$ solar cell with a 1.4μm S-graded buffer, for the following values of dislocation density: (a) $D=0$ cm^{-2}. (b) $D=10^5$ cm^{-2}. (c) $D=10^6$ cm^{-2}. (d) $D=10^7$ cm^{-2}. (e) $D=10^8$ cm^{-2}. (f) $D=10^9$ cm^{-2}.

An interesting observation is that the dislocation density at the top and bottom of the S-graded layer may also be estimated from the widths of the two peaks associated with this type of buffer. However, the linearly-graded buffer exhibits only a single broad band of diffraction so it appears that a similar type of analysis will be difficult to apply in that case.

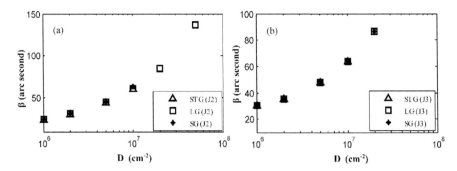

Fig. 8. The widths of 004 rocking curves for step-graded (STG, open triangles), linearly-graded (LG, open squares), and S-graded (SG, crosses) triple-junction solar cells as a function of the threading dislocation density. (a) middle junction (J2); (b) top junction (J3).

Next we set out to determine the correlation between the observed width and the thread density. Figure 8 shows the widths of 004 diffraction peaks from the middle and top junctions in step-graded, linearly-graded, and S-graded triple-junction solar cells as a function of the dislocation density, assumed to be uniform throughout the structure. All peak widths show a similar monotonic dependence on the dislocation density, so it will be possible to estimate the dislocation densities by straightforward application of the results in this figure. It should be noted, however, that closer lattice matching of the top two junctions will make independent determination of the two dislocation densities difficult, because a single peak may be observed, and its width may be determined in part by the lattice mismatch as well as the dislocation densities in the individual layers. In the case of exact lattice matching, the weighted average dislocation density for the two top junctions may be determined, even though their densities may not be found independently. This weighted average density would still provide important feedback to the crystal grower, and could be used to compare different approaches to the graded buffer layers.

In Fig. 9 the 004 rocking curve widths for the middle junction are compared to the calculations of the convolution model, for step-graded, linearly-graded, and S-graded triple-junction solar cells, as a function of the threading dislocation density. The widths determined by the convolution model have been correlated with experimentally-determined rocking curve widths for single uniform layers with known dislocation densities, and predicts a rocking curve width given by [23-24]

$$\beta^2 \approx \beta_0^2 + \beta_\alpha^2 + \beta_\beta^2$$
$$= \beta_0^2 + 2\pi b^2 D \cdot \ln(2) + 0.16\, b^2 D \left|\ln(2\times 10^7 cm\sqrt{D})\right| \tan^2(\theta_B) \quad (19)$$

where β_α and β_β respectively are the widths of the angle-scale distributions associated with angular mosaic spread and strain mosaic spread. β_o is the intrinsic rocking curve width determined by using a dynamical simulation for a perfect crystal with the same thickness. Figure 9 shows that there is good agreement between the 004 rocking curve widths from the calculated diffraction profiles and the predictions of the convolution model for the

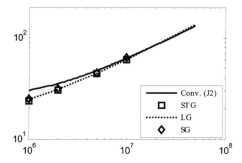

Fig. 9. The rocking curve widths for the middle junction (J2) observed in 004 rocking curves for step-graded (STG, open squares), linearly-graded (LG, dotted line), and S-graded (SG, open diamonds) triple-junction solar cells as a function of the threading dislocation density, and compared with the convolution model (Conv., solid).

middle junction (J2). Therefore the convolution model analysis may be applied to estimate the dislocation density in the middle junction. A similar conclusion holds for the top junction (J3), as can be seen from the results contained in Fig. 8, as long as the diffraction peaks from the middle and top cells are resolved.

Similar conclusions hold for the 044 and 117 reflections as well. Figures 10-12 illustrate the 044 rocking curves, and Figs. 13-15 the 117 rocking curves, for structures with the three types of buffer layers, and for the following values of the dislocation density: 0, 10^5, 10^6, 10^7, 10^8, and 10^9 cm^{-2}. The rocking curve peak widths for the top two junctions increase monotonically with the dislocation density, and the defect density may be estimated using the convolution theory up to densities of $\sim 10^8$ cm^{-2}, beyond which peak overlap complicates the analysis and detailed mosaic crystal calculations should be applied instead. The higher-order reflections result in a greater angular range for significant rocking curve intensity and also cause increased broadening of the individual peaks at a given dislocation density. There is a complex interplay of these two factors, with the net result that the 044 reflection is most sensitive to the dislocation density in these structures. Figure 16 summarizes these results for the top junction, and also shows additional results for the 113, 115, and 026 reflections, providing the dependence of the rocking curve width on dislocation density for the S-graded structure, and demonstrating agreement with the convolution approximation.

In practical metamorphic structures, the threading dislocation density may not be constant, but typically decreases with distance from the interface due to annihilation and coalescence reactions between threading defects. We therefore calculated 004 rocking curves for S-graded triple junction solar cell structures in which the dislocation density was assumed to decrease linearly from 10^7 cm^{-2} at the substrate interface to 10^6 cm^{-2} at the surface (Fig. 17(a)). For comparison, we also calculated the 004 rocking curve for the same structure in which the dislocation density was assumed to increase linearly from 10^6 cm^{-2} at the substrate interface to 10^7 cm^{-2} at the surface (Fig. 17(b)). In the former case, the higher-angle feature from the S-graded buffer (associated with the bottom portion of the buffer) is broadened to a greater extent than the lower-angle feature (associated with the top portion of the buffer), whereas in the latter case the opposite is seen to be true. Quantitative analysis of the broadening therefore should allow estimation of the dislocation density for the top and bottom of the S-graded buffer layer, as we show next.

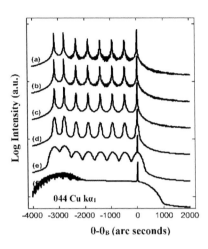

Fig. 10. The 044 rocking curves for a $Ga_{0.33}In_{0.67}P/In_xGa_{1-x}As/Ge$ solar cell with a step-graded buffer (7 steps, 200 nm each) for the following values of dislocation density: (a) D=0 cm^{-2}. (b) D=10^5 cm^{-2}. (c) D=10^6 cm^{-2}. (d) D=10^7 cm^{-2}. (e) D=10^8 cm^{-2}. (f) D=10^9 cm^{-2}.

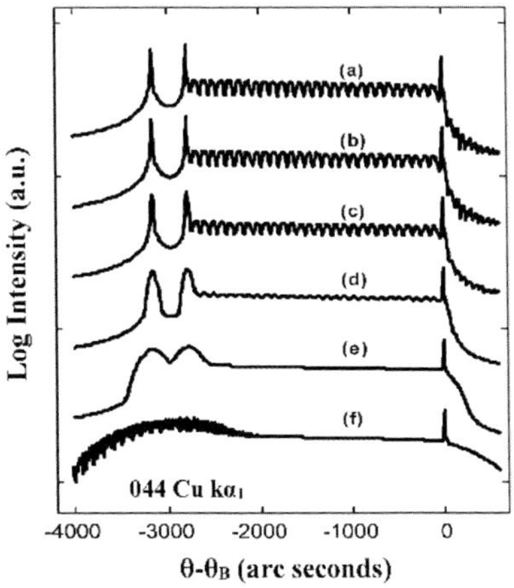

Fig. 11. The 044 rocking curves for a $Ga_{0.33}In_{0.67}P/In_xGa_{1-x}As/Ge$ solar cell with a 1.4μm linearly-graded buffer, for the following dislocation densities: (a) D= 0 cm^{-2}. (b) D=10^5 cm^{-2}. (c) D=10^6 cm^{-2}. (d) D=10^7 cm^{-2}. (e) D=10^8 cm^{-2}. (f) D=10^9 cm^{-2}.

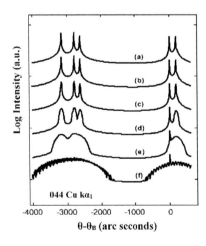

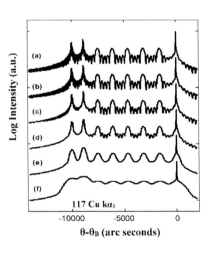

Fig. 12. The 044 rocking curves for a $Ga_{0.33}In_{0.67}P/In_xGa_{1-x}As/Ge$ solar cell with a 1.4μm S-graded buffer layer, for the following dislocation densities: (a) D=0 cm^{-2}. (b) D=10^5 cm^{-2}. (c) D=10^6 cm^{-2}. (d) D=10^7 cm^{-2}. (e) D=10^8 cm^{-2}. (f) D=10^9 cm^{-2}.

Fig. 13. The 117 rocking curves for a $Ga_{0.33}In_{0.67}P/In_xGa_{1-x}As/Ge$ solar cell with a step-graded buffer layer (7 steps, 200 nm each), for the following dislocation densities: (a) D=0 cm^{-2}. (b) D=10^5 cm^{-2}. (c) D=10^6 cm^{-2}. (d) D=10^7 cm^{-2}. (e) D=10^8 cm^{-2}. (f) D=10^9 cm^{-2}.

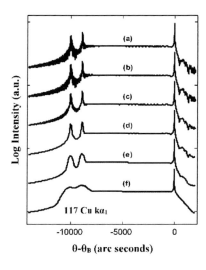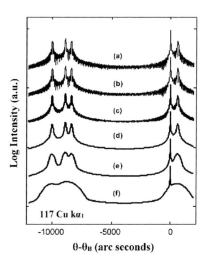

Fig. 14. The 117 rocking curves for a $Ga_{0.33}In_{0.67}P/$ $In_xGa_{1-x}As/$ Ge solar cell with a 1.4μm linearly-graded buffer layer, for the following dislocation densities: (a) D=0 cm^{-2}. (b) D=10^5 cm^{-2}. (c) D=10^6 cm^{-2}. (d) D=10^7 cm^{-2}. (e) D=10^8 cm^{-2}. (f) D=10^9 cm^{-2}.

Fig. 15. The 117 rocking curves for a $Ga_{0.33}In_{0.67}P/$ $In_xGa_{1-x}As/Ge$ solar cell with a 1.4μm S-graded buffer layer, for the following dislocation densities: (a) D=0 cm^{-2}. (b) D=10^5 cm^{-2}. (c) D=10^6 cm^{-2}. (d) D=10^7 cm^{-2}. (e) D=10^8 cm^{-2}. (f) D=10^9 cm^{-2}.

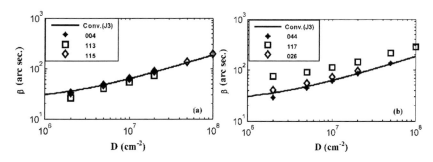

Fig. 16. Rocking curve widths for the top junction (J3) of a $Ga_{0.33}In_{0.67}P/In_xGa_{1-x}As/Ge$ solar cell with a 1.4μm S-graded buffer layer as a function of the dislocation density, and compared to the calculation of the convolution model for the case of the 004 reflection. (a) 004, 113 and 115 reflections; (b) 044, 117 and 026 reflections.

We have shown previously that the dislocation density in a step-graded layer may be estimated from the widths of the individual peaks using either the mosaic crystal model or the convolution approximation, as long as the individual peaks may be resolved [28]. Such an investigation has never been conducted for S-graded layers, however. Here we sought to determine if the dislocation density in the S-graded layer could be estimated in a simple way from the widths of the two peaks corresponding to the top and bottom regions of the S-graded buffer. Figure 18 shows the widths of these two peaks from the S-graded buffer layer as a function of the threading dislocation density (assumed to be uniform) and compares these widths to the convolution approximation. It can be seen that the two widths

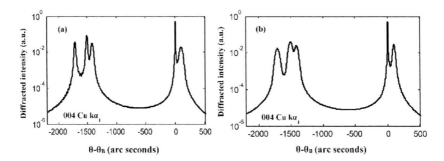

Fig. 17. The 004 rocking curves for a $Ga_{0.33}In_{0.67}P/In_xGa_{1-x}As/Ge$ solar cell with a 1.4μm S-graded buffer. (a) The dislocation density is assumed to decrease linearly from 10^7 cm^{-2} at the substrate interface to 10^6 cm^{-2} at the surface; (b) the dislocation density is assumed to increase linearly from 10^6 cm^{-2} at the substrate interface to 10^7 cm^{-2} at the surface.

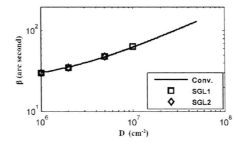

Fig. 18. The 004 rocking curve widths of the peaks associated with the S-graded buffer in a $Ga_{0.33}In_{0.67}P/In_xGa_{1-x}As/$ Ge solar cell as a function of the threading dislocation density. The predicted peak width for the layer adjacent to the interface (SGL1, open squares) and the layer adjacent to the middle junction (SGL2, open diamonds) are compared to the calculations of the convolution model (Conv, solid line).

exhibit the same monotonic variation with the dislocation density, and that this variation is consistent with the convolution model. Therefore the thread density in the S-graded buffer may be estimated simply by application of the convolution model to either peak, as long as one of the peaks can be resolved in the structure of interest.

4. Conclusion

We described the analysis of the threading dislocation densities for step-graded, linearly-graded and S-graded triple junction metamorphic solar cells grown on Ge (001) substrates. Analysis based on the convolution approximation may be used to estimate the dislocation densities in each of the solar cell junctions, the step-graded buffer, and the S-graded buffer, as long as significant peak overlap is absent. Often practical triple-junction cells of this type are fabricated with the top two junctions lattice-matched to one another. In such a cell, the threading dislocation densities in the top cells could not be determined independently, because of the overlap of their diffraction peaks. However, the weighted average of the dislocation densities in the top junctions could be determined from the width of the single observed peak, and this would still provide valuable feedback to the crystal grower.

Acknowledgement

The work has been supported in part by Taif University and Epitax Engineering. This support is gratefully acknowledged.

References

1. S. Takagi, Acta Cryst., 15, 1311 (1962).
2. D. Taupin, Bull. Soc. Fr. Mineral. Cristallogr. 87, 469–511 (1964).
3. S. Takagi, J. Phys. Soc. Jpn., 26, 1239 (1969).
4. M. A. G. Halliwell, M. H. Lyons, and M. J. Hill, J. Cryst. Growth, 68, 523 (1984).
5. C. R. Wie, T. A. Tombrello, and T. Vreeland, Jr., J. Appl. Phys., 59, 3743 (1986).
6. W. J. Bartels, J. Hornstra, and D. J. W. Lobeek, Acta Cryst., A42, 539 (1986).
7. V. S. Speriosu, J. Appl. Phys., 52, 6094 (1981).
8. V. S. Speriosu and T. Vreeland, Jr., J. Appl. Phys., 56, 1591 (1984).
9. L. Tapfer and K. Ploog, Phys. Rev. B, 40, 9802 (1989).
10. C. R. Wie, J. Appl. Phys., 65, 1036 (1989).
11. C. R. Wie and H. M. Kim, J. Appl. Phys., 69, 6406 (1991).
12. P. B. Rago, E. N. Suarez, F. C. Jain, and J. E. Ayers, J. Electron. Mater., doi: 10.1007/s11664-012-2012-y (2012).
13. P. B. Rago and J. E. Ayers, J. Electron. Mater., 42, 2450-2458 (2013).
14. P. B. Rago, J. E. Raphael, and J. E. Ayers, Mater. Sci. Technol. Conf., Montreal, PQ, 27–31 October 2013.
15. P. B. Rago, F. C. Jain, and J. E. Ayers, J. Electron. Mater., 42, 3066-3070 (2013).
16. P. B. Rago and J. E. Ayers, J. Vac. Sci. Technol. B, 33, 021204 (2015).
17. S. Xhurxhi, F. Obst, D. Sidoti, B. Bertoli, T. Kujofsa, S. Cheruku, J. P. Correa, P. B. Rago, E. N. Suarez, F. C. Jain, and J. E. Ayers, J. Electron. Mater., 40, 2348 (2011).
18. T. Kujofsa, A. Antony, S. Xhurxhi, F. Obst, D. Sidoti, B. Bertoli, S. Cheruku, J. P. Correa, P. B. Rago, E. N. Suarez, F. C. Jain, and J. E. Ayers, J. Electron. Mater., 42, 3408 (2013).
19. T. Kujofsa and J. E. Ayers, Intl. J. High Speed Electron. Sys., 23, doi: 10.1142/S0129156414200055 (2014).
20. J. A. Prins, Z. Phys., 63, 477 (1930).
21. P. Gay, P. B. Hirsch, and A. Kelly, Acta Met., 1, 315 (1953).
22. M. J. Hordon and B. L. Averbach, Acta Met., 9, 237 (1961).
23. J. E. Ayers, J. Cryst. Growth, 135, 71 (1994).
24. P. D. Healey, K. Bao, M. Gokhale, J. E. Ayers, and F. C. Jain, Acta. Cryst. Sec. A, A51, 498 (1995).
25. W. Guter, J. Schone, S. P. Philipps, M. Steiner, G. Siefer, A. Wekkeli, E. Welser, E. Oliva, A. W. Bett, and F. Dimroth, J. Appl. Phys. Lett., 94, 223504 (2009); doi: 10.1063/1.3148341.
26. W. J. Bartels, J. Vac. Sci. Technol. B, 1, 338 (1983).
27. P. D. Healey and J. E. Ayers, Acta Cryst., A52, 245 (1996).
    F. Althowibi, P. B. Rago, and J. E. Ayers, J. Vac. Sci. Technol. B, in press (2016).

Pixel Characterization of a Protein-Based Retinal Implant Using a Microfabricated Sensor Array

Jordan A. Greco[†,¶], Luis André L. Fernandes[‡], Nicole L. Wagner[†],
Mehdi Azadmehr[‡], Philipp Häfliger[§], Erik A. Johannessen[‡] and Robert R. Birge[†,ǁ]

[†]*Department of Chemistry,*
University of Connecticut, 55 North Eagleville Road, Storrs, CT 06269, USA
[‡]*Department of Microsystems (IMS),*
University College of Southeast Norway, Campus Vestfold,
Raveien 215, 3184 Borre, Norway
[§]*Department of Informatics (IFI),*
University of Oslo, P.O Box 1080 Blindern, 0316 OSLO, Norway
[¶]*jordongreco9@gmail.com*
[ǁ]*rbirge@uconn.edu*

Retinal degenerative diseases are characterized by the loss of photoreceptor cells within the retina and affect 30-50 million people worldwide. Despite the availability of treatments that slow the progression of degeneration, affected patients will go blind. Thus, there is a significant need for a prosthetic that is capable of restoring functional vision for these patients. The protein-based retinal implant offers a high-resolution option for replacing the function of diseased photoreceptor cells by interfacing with the underlying retinal tissue, stimulating the remaining neural network, and transmitting this signal to the brain. The retinal implant uses the photoactive protein, bacteriorhodopsin, to generate an ion gradient in the subretinal space that is capable of activating the remaining bipolar and ganglion cells within the retina. Bacteriorhodopsin can also be photochemically driven to an active (**bR**) or inactive (**Q**) state, and we aim to exploit this photochemistry to mediate the activity of pixels within the retinal implant. In this study, we made use of a novel retinomorphic foveated image sensor to characterize the formation of active and inactive pixels within a protein-based retinal implant, and have measured a significant difference between the output frequencies associated with the **bR** and **Q** states.

Keywords: protein-based retinal implants; foveated image sensors; bacteriorhodopsin; Q state; pixel mediation.

1. Introduction

Retinal degenerative diseases, including retinitis pigmentosa (RP) and age-related macular degeneration (AMD), involve the irreversible degeneration of the retinal photoreceptor cells.[1,2] These diseases cause the layer of light sensing cells in the eye to degenerate over time, however, a significant portion of the underlying retinal tissue that transmits visual information to the brain remains largely intact.[2-4] The resulting loss of

[ǁ]Corresponding author.

vision causes a significant decline in the quality of life for those affected. Retinitis pigmentosa, the most common of the inherited retinal degenerative disorders, affects approximately 1.5 million people worldwide and can affect people of all ages.[5,6] Age-related macular degeneration is the leading cause of irreversible blindness among the elderly and is estimated to affect approximately 30 million people globally.[7,8] There are currently no cures for patients with RP or AMD, and all available treatments only slow progression, are limited in effectiveness, and ultimately fail to prevent permanent loss of vision. Thus, there is a significant need for a therapy or prosthetic capable of restoring functional vision to RP and AMD patients.

Many new biotechnologies and treatments for blindness resulting from retinal degeneration have been developed during the past decade. While these treatments are still in their infancy and are limited by inherent drawbacks, a number of diverse approaches are currently under investigation in order to achieve a cure for diseases like RP and AMD. Novel and emerging approaches to treat retinal degeneration include stem cell tissue transplants,[9-14] gene therapies,[15-17] and optogenetic technologies.[18-24] Despite the promise of these treatments in design and preliminary efficacy, these biotechnologies are still under development and are now undergoing early clinical trials. Currently, intra-ocular and periocular injections and nutritional drug supplements are treatments which have shown success in slowing the progression of degeneration, however, limited efficacy is observed and the treated patient will eventually go blind.[8,25,26] Electrode-based retinal implants, which currently represent the most common type of prosthetic used to restore vision, are being designed, developed, and commercialized in an effort to replace the function of the damaged photoreceptor cell layer within the retina. These retinal implants generally utilize an external apparatus to capture an image, convert that image into electrical signals, and stimulate the remaining neural circuitry within the degenerated retina. It has long been demonstrated that electrical stimuli can initiate the visual cascade when delivered in the extracellular domain of a retinal neuron.[27] Retinal implant architectures that exploit this electrical stimulation have seen some promise, and a number of companies and academic groups have demonstrated efficacy in the clinical setting, including efforts by Second Sight,[28-30] Retinal Implant AG,[31-35] and the Epi-Ret 3 team.[36-38]

Of particular significance is the technology developed by Second Sight, which continues to be at the forefront of retinal prosthetic development and commercialization. The Argus II, created by Second Sight,[28-30] was the first approved retinal prosthetic for clinical trials in both the US and Europe, and is currently the only approved implant on the market. The design of the electrode-based implant consists of a 60-electrode array placed in an epiretinal position. The Argus II has been implanted in over 100 patients with RP, and the patients were capable of detecting some motion and performing simple mobility tasks.[30] However, these retinal prosthetics have intrinsic shortcomings.[39] First, replicating the spatiotemporal patterns of the neurosensory network with electrode arrays is a challenge.[40] This problem is exacerbated by the low resolution associated with the electrode arrays, resulting in a limited ability to detect simple direction of motion through high contrast, black and white gradients.[28,41] Secondly, because the implant contains transscleral cables that penetrate the eye, the risk of infection is high and a number of serious adverse effects were observed in as many as 9 out of 30 patients, including conjunctival erosion, endophthalmitis, and hypotony.[29,42] Furthermore, implantation requires laborious, complex surgery, and thereby limits adoption by the vitreoretinal surgeon community.[39,43] According to lead researchers in the field, there is still a significant need for a device of higher resolution that also mitigates the risk of surgical complications.

The protein-based retinal implant offers unique solutions to the inherent shortcomings demonstrated in current electrode-based retinal prosthetics.[44] The retinal implant architecture is manufactured *via* a bottom-up approach and is comprised of multiple layers of the light-activated proton pump, bacteriorhodopsin (BR), oriented on a flexible, ion-permeable membrane *via* layer-by-layer (LBL) electrostatic adsorption.[45] The high resolution, subretinal implant converts light energy into an ion gradient that is capable of stimulating the remaining neural circuitry of the degenerated retina (Fig. 1).[44] Like the electrode-based technologies, the protein-based prosthetic relies on the presence of the inner retinal tissue (i.e., intact bipolar and ganglion cells) in order to convert absorbed light energy into an electrochemical gradient that is interpreted by the brain as meaningful visual perception. The small and flexible protein-based retinal implant is a standalone prosthetic, which is capable of responding only to incident light without the influence of external devices that penetrate or communicate through the ocular tissue. The optical resolution of the implant is comparable to that of native photoreceptors due to the nanometer-scale features of the protein and the molecular packing of the biomaterials within the implant structure. Moreover, the unique photochemistry of the protein can be used to calibrate the retinal implant by modulating the active pixel area relative to the extent of retinal degeneration.[46]

The protein-based retinal implant architecture harnesses the inherent pumping capability of BR, a feature necessary for the survival of the native organism, to generate a macroscopic ion gradient for retinal stimulation. Bacteriorhodopsin is a 26-kDa photoactive protein found in the outer membrane of the halophilic *archaeon*, *Halobacterium salinarum*.[47,48] When the concentration of oxygen is insufficient to sustain growth *via* respiration (i.e., ATP generation by oxidative phosphorylation),[47,48] BR is expressed within a two-dimensional crystalline lattice of trimers, known as the purple membrane.[49,50] Upon the absorption of a photon by the protein-bound chromophore, all-*trans* retinal, BR transports a proton from the intracellular domain to the extracellular milieu *via* a series of transient photochemical intermediates, known as the photocycle (Fig. 2).[51] The net translocation of a proton generates a proton gradient, thereby driving cellular ATPase to synthesize ATP under anaerobic conditions.[47]

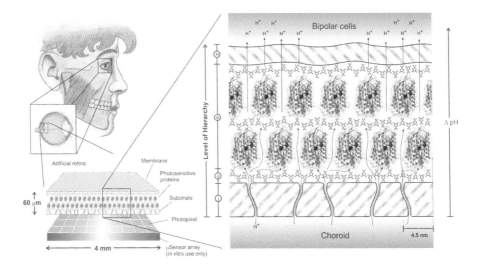

Fig. 1. Architecture and application of the proposed protein-based retinal implant (or artificial retina). The implant acts as a subretinal implant located below the retinal tissue in place of degenerated photoreceptor cells. The implant consists of two ion-permeable membranes, depicted as (i) and (iv) in the RHS close up, with multilayers of the photosensitive protein, bacteriorhodopsin, contained between them. A polycation (ii) permits the deposition of alternating layers of the protein and polycation (iii) through a layer-by-layer electrostatic adsorption process. The orientation of the implant permits the generation of an H$^+$ gradient towards the remaining neural network of the retina. This novel approach of stimulating the intact bipolar cells mimics the native phototransduction cascade initiated by healthy photoreceptor cells. A photopixel microsensor array (bottom LHS image) was used in this study to examine the photoactive state of the bacteriorhodopsin *in vitro*, but is not a part of the prosthetic.

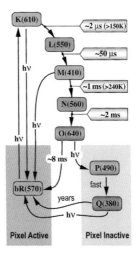

Fig. 2. The main and branched photocycles of light adapted BR (**bR**). Upon the absorption of a single photon, the protein cycles through a series of transient intermediates (**K**, **L**, **M**, **N**, and **O**) and subsequently returns to the **bR** resting state. In order to access the branched photocycle (the **P** and **Q** states), a second photon must be absorbed to photochemically convert the **O** state to **P**. The resulting **Q** state is stable for several years at physiological temperature. The purple and yellow boxes also highlight the states in which the pixels within a retinal implant would be active or inactive, respectively. This figure is based on Fig. 1 from Ref. 52.

The proton pumping mechanism initiated by the absorption of a photon by the retinal chromophore of BR is one of the most efficient photochemical reactions found in nature. The high quantum efficiency of this reaction (~65%) is identical to that of the visual pigment, rhodopsin, which is found in the photoreceptor cells of the human retina.[53,54] Moreover, BR is a natural candidate for biophotonic devices due to a remarkable thermal stability and photochemical efficiency. Bacteriorhodopsin has a melting temperature in excess of 80 °C,[55] and has an observed photochemical cyclicity that exceeds 10^6 photocycles before a 37% loss of the irradiated ensemble.[56-58] Because BR is capable of withstanding high fluctuations in temperature, light flux, and chemical stress from a self-induced pH gradient, the biomaterial has been implemented as the photoactive medium in a number of devices for several decades.[52,59]

Figure 2 provides a schematic of the BR photocycle and also highlights the ability of converting active pixels in a protein-based retinal implant to inactive pixels. This process is facilitated by accessing the branched photocycle (the **P** and **Q** states) *via* a sequential multiphoton process. The **Q** state, which was first described by Popp *et al.*,[60] contains a hydrolyzed 9-*cis* chromophore and differs from the other photointermediates within the BR photocycle because it is a thermally stable photoproduct that could last several years at ambient temperature.[52,61] The **Q** state can be driven back to the resting (**bR**) state using high intensity UV light.[62] Despite the observation of this photochemistry, access to the branched photocycle is highly inefficient for the native protein. Consequently, a significant research effort has been undertaken to use genetic engineering and directed evolution to identify enhanced mutants with efficient **Q** state formation.[46,63] Over 10,000 mutants were screened during this study, and a number of mutants emerged with improved **Q** state performance, including the mutant V49A.[46] The enhanced BR biomaterial is capable of forming two stable and spectrally discrete photoproducts (**bR** and **Q**), which have been exploited as differentiable bits (bit 0 and bit 1) in protein-based volumetric memories and processors.[52,64-66] Similarly, this photochemistry can provide a means to mediate pixels on the active surface area of a protein-based retinal implant. Because the **Q** state would prevent the protein from pumping protons, it has the potential to deactivate specific areas of the implant to prevent interference with functional photoreceptor cells following surgical implantation.

In this study, we seek to characterize the ability to mediate pixels within a protein-based retinal implant comprised of the high **Q**-forming mutant, V49A. Our approach makes use of a foveated complementary metal-oxide semiconductor (CMOS) imager, which is based on neural-inspired camera technologies that detect and process light similarly to the human eye. A foveated imager is a sensor that combines different pixel properties, which are dependent on their measured distance or computational delay relative to the center.[67] In using higher resolution in the center of the chip, one can simulate the fovea (higher concentration of cone photoreceptors) while keeping larger pixels towards the periphery of the chip working as large dynamic motion detectors (equivalent to the rods in the retina). Because we are not measuring the response to movement, we will focus on the properties of the static pixels only, leaving the use of the larger dynamic peripheral pixels for future studies. The high resolution, sensitivity, and low noise properties of CMOS image sensors make them very good candidates for test platforms that can simulate the human retina and test the photosensitive properties of novel BR-based retinal implants.

Protein-based retinal implants were generated using the V49A BR mutant, and the implants were driven to the **Q** state by using LED-induced photochemistry. The hydrated retinal implants were then placed onto the foveated image sensor for relative light intensity measurements by the photodiodes, which is translated into an output frequency consisting of a series of action potentials (APs). At the conclusion of this study, we have shown that the image pixel properties can be extracted by attributing a light intensity to each of the pixels, which will allow for the differentiation between the **bR** or **Q** states and provide a basis for quantifying the extent to which the **Q** state has formed through a pixel light intensity map. The results of these proof-of-concept experiments show preliminary characterization of pixel mediation within a protein-based retinal implant, and ongoing experiments aim to quantify the limits of spatial resolution and generate the ability to drive local pixels between the two photoproducts.

2. Methods and Materials

2.1. *Chemicals and buffers*

All chemicals were purchased from Fisher Scientific, Inc. (Pittsburg, PA) or Sigma Aldrich (St. Louis, MO). Because an alkaline pH is necessary to produce and isolate the **Q** photoproduct, a 50 mM glycine buffer at pH 9.5 was prepared and used for all solution-based experiments in this study.[46]

2.2. *Strain generation, protein isolation, and purification*

The high **Q**-forming mutant, V49A, was first identified using Type I directed evolution.[46] In order to express this mutant form of BR, mutant DNA was transformed into the MPK409 cell line of *H. salinarum* using methods outlined by Peck *et al.*[68] Purple membrane fractions were then prepared and isolated according to standard procedures.[69,70]

2.3. *Manufacturing of the protein-based retinal implants*

The protein-based retinal implants were comprised of a multilayered, BR-based thin film generated *via* sequential electrostatic adsorption, which was achieved through a LBL manufacturing technique.[45,71,72] The multilayered thin films are capable of harnessing and amplifying the proton pumping action of BR if the protein is uniformly oriented in the film at an optimal optical density. There must be enough layers of BR to adequately absorb incident light while also generating an appreciable unidirectional ion gradient for retinal stimulation. The solid support surface of the thin film is a bioinert, ion-permeable mesh comprised of polyethylene terephthalate (PET) microfibers. This material has previously found success for use in the eye[73,74] and is amenable to surface modification in order to serve as a scaffolding for multilayers of BR within our protein-based retinal implant. Because the LBL process requires a charged surface for subsequent electrostatic adsorption, the PET film was first exposed to conditions that facilitate the reduction of surface carbonyl functional groups that renders the surface negatively charged.[75-77]

Following preparation of the PET scaffolding, the LBL manufacturing technique was implemented as first described by He *et al*.[45,71,72] In brief, the PET film is first dipped in a solution of poly(diallyldimethyl ammonium chloride), followed by rinse periods in ddH$_2$O (Millipore, Billerica, MA) and a short drying interval. The thin film is then transferred to a solution of BR and rinsed twice in basic glycine buffer. It is important to note that this technique was carried out so that only one surface of the film was coated during the LBL methods outlined above.

2.4. Preparation of the Q state

The bacteriorhodopsin mutant used in this study (V49A) was selected due to the identification of this protein as a high **Q**-forming mutant.[46] The **Q** photoproduct was formed while the protein was contained within the multilayered thin film. The thin films were placed in a petri dish filled with enough glycine buffer (50 mM; pH 9.5) to completely hydrate the film for the duration of the illumination period. The film was placed under an LED light bank (100 mW/cm^2), which contained 12 red (> 640 nm) Luxeon III Lambertian LEDs that were each driven at 850 mA, for 8 hours at ambient temperature. Prior to further experiments, the **Q** state thin film was hydrated within the buffered solution and the petri dish was wrapped in aluminum foil to prevent exposure to ambient light.

2.5. Absorption spectroscopy

All absorption spectra were collected at ambient temperature using a Varian Cary 5000 UV-visible spectrophotometer (Palo Alta, CA). The retinal implants were inserted into a 1 mm quartz spectrophotometer cell (Starna Cells, Inc.; Atascadero, CA) and the films were suspended in 50 mM glycine buffer (pH 9.5). A bare PET mesh film in buffer was used as the blank for all measurements.

2.6. Foveated image sensors

Modern CMOS sensor-based cameras offer an excellent optical resolution.[78] The CMOS image sensors that form the core of these cameras benefit from technology scaling that has provided them with a resolution increase (pixel size reduction) combined with the possibility to include more on-chip image processing circuitry. However, size scaling beyond sub-micron (< 0.25 μm) processing technologies is limited by dark current noises, tunneling effects through thin gate oxides, and a low photoresponsivity due to a combination of shallow junctions and high doping.[79] Continuous effort is seeking to negotiate these limiting parameters through a gradual improvement of the fabrication technologies used, which together with the low cost of manufacture have continued to secure the commercial success of CMOS image sensors. This has, in turn, paved the way for new application possibilities, such as neural-inspired or retinomorphic cameras,[80,81] that mimic the way human eye detects and processes light and images.

The foveated image sensor used in this project was created using the commercial 0.35 μm CMOS processing technology, which measures 3.15 × 3.15 mm² in extent.[67] Like most neural-inspired circuits, this camera communicates with voltage pulses that resemble APs. Each pixel consists of a photodiode and the associated processing circuitry. The circuitry of the static photopixel cells (located in the center of the image sensor) consists mainly of an integrator. It integrates a current linearly proportional to the light level over a short period of time, and when a threshold is reached, an AP is fired. The AP is conveyed off-chip by peripheral logic and the pixel is reset (Fig. 3). Thus, the firing frequency reflects the light intensity collected by this pixel.

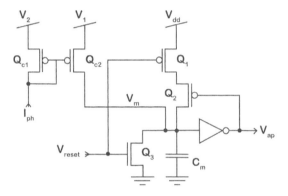

Fig. 3. Foveated imager pixel logic. The photodiode is connected to I_{ph} and the current is mirrored in Q_{c1}-Q_{c2}, which can also act as an amplifier (tilted current mirror), charging C_m. Once V_m has reached the inverter threshold, it spikes an AP (V_{ap}) and initiates the communication circuit. Q_2 opens, and since V_{reset} is active high, Q_1 is already open and V_m is pulled to V_{dd}. When the communication circuit sends a reset signal to the pixel, V_m is set to ground and the reset signal goes low again, charging C_m once more.

In more detail, the photo current i conveyed by the current mirror charges a capacitor C_m and, when the voltage across it reaches the threshold of the inverter, the output from the inverter changes from state 1 to 0 (digital representation). Through a feedback loop, the transistor Q_2 is activated and capacitor C_m is stabilized at the potential V_{dd} when the reset signal is low. When the reset signal is asserted, the transistor Q_1 will deactivate, and C_m will empty its charge to ground through Q_3, reaching the initial state. The size of C_m is essential to determine how fast the threshold voltage of the inverter will be reached. The voltage at C_m (V_m) is given by Eq. (1):

$$V(t)_m = \frac{1}{C} \int_0^t i d\tau + V(0). \tag{1}$$

For $V(0) = 0$:

$$V(t)_m = \frac{1}{C}\int_0^t i\,d\tau. \qquad (2)$$

From Eq. (2), it is clear that the use of a small capacitor (capacitance, C) will reach the threshold potential ($V(t)_m$) faster. This also comes at the cost of a noisier circuit in which the leakage current (dark current at zero light intensity) charges a smaller capacitor faster, resulting in the generation of APs. In contrast, a larger capacitor will take a longer time to charge under these conditions, thus deceasing the rate of APs and, consequently, the noise. This makes it less dependent on the dark current, but is at the cost of operating as a slower circuit. The area of the capacitor used in the static pixels of the foveated image sensor, $C_m = 2.48 \times 10^{-4}$ µm^2, corresponds to a capacitance of 221.58 fF. The dark current will charge this capacitor at a rate that corresponds to the generation of 1 AP/s at zero lux. Increasing the light level to 3 lux (comparable to the test conditions described below) increases the firing rate to 8 AP/s. Hence, a change in the photocycle state of BR will reflect a change in light intensity and, consequently, a shift in the AP frequency.

The dynamic photopixel cells that are located in the periphery of the chip fire an AP if changes in light are detected according to the photoreceptor circuit described by Delbrück and Mead,[82] where an electronic photoreceptor outputs a voltage that is logarithmic to the light intensity. The dynamic pixels are adaptive to various light levels and the APs are fired as the speed of that adaptation exceeds a certain threshold.[67] The photodiodes in all of the pixels are made from substrate diodes (n-well regions defined by, e.g., boron doping) in the bulk p- substrate of the chip.

Fig. 4. Photomicrograph of the foveated image sensor. Four rows of dynamic pixels (616 total pixels) surround the central 76 × 67 array of static pixels.

While still large when compared to commercial CMOS image sensor pixels due to the in-pixel logic, the static photopixels are smaller than the dynamic pixels in order to produce high resolution pictures, measuring 26.8 × 25.2 µm² with a fill factor (photodiode area) of 11.3%. The dynamic pixels are larger, measuring 53.6 × 50.4 µm², with a fill factor of 50.4%. The photodiodes for the dynamic pixels are larger in order to achieve a higher sensitivity to light intensity variations. The pixel array is made up of 76 × 67 static cells in the center surrounded by 616 dynamic pixels according to (Fig. 4).

2.6.1. AER communication protocol

The schematic in Fig. 3 (except for the current mirror) is called an Integrate and Fire circuit and is a good example of a mixed-mode circuit. It is based on the "self-reset neuron".[83] A real biological neural network, which consists of point-to-point connections, is difficult to implement on an artificial neural CMOS chip due to a limiting amount of metal layers that are available for routing combined with a limited amount of contact pads that are available for interconnecting the generated signals to an external circuitry. A solution for asynchronous neuromorphic communication is the Address Event Representation (AER) protocol.[84] Most neurons communicate by way of nerve pulses or APs *via* dedicated point-to-point connections (axons). This is in contrast to the communication channels between computers, or inside computers, that transmit more complex signals at higher rates than an axon. The physical channels are mostly shared and time multiplexed by several components keeping a lower density. As neuromorphic engineers try to implement systems that are organized more like the nervous system, communicating internal signals like the brain can become a major obstacle to electronics. The human brain contains about 10^{11} neurons, each of which has 1,000 to 10,000 synapses. All of those synapses receive input *via* a dedicated output line of the sending neuron, making the brain densely packed with connections between cortical areas and other parts.

Since neurons/pixels use APs to communicate, the AER bus is used to send the address of the neuron/pixel off-chip, either to a receiving neuromorphic circuit or a computer where the activity can be visualized, e.g., as a gray scale image of average pixel activity. This communication protocol takes advantage of the speed in integrated circuits, compensating for the lack of connections. The functioning principle can be explained as follows. When a neuron wishes to send an AP, it places its address on a digital bus *via* an encoder. Synapses that are supposed to receive that AP are connected to the same bus *via* a decoder and get stimulated when their address occurs (Fig. 5). The bus can only transmit APs serially but does it much faster than an axon. It can transmit APs so tightly spaced in time that, for a network of neurons that operate on a natural time scale, these pulses are virtually instantaneous. Since multiple point-to-point connections share the same bus, it is necessary to have a bus control mechanism for handling collisions. The AER communication protocol is described in detail in Ref. 67.

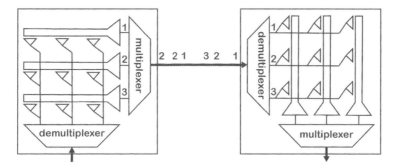

Fig. 5. Schematic of the Address Event Representation (AER) principle.

2.7. Q state detection

In order to characterize the pixel content contained within the protein-based retinal implant, each film was placed directly on top of the foveated image sensor with a particular emphasis on the static pixels within the photodiode array. The retinal implants were either in the **bR** resting state or driven to the **Q** state using the photonic criteria described above.[46] Each film was placed in a consistent orientation on the chip so that the pixel intensity measurements were under identical conditions. The sensor array was embedded in a sensor housing that protected the electrical interconnections from the hydrated films. Silicone purchased from Dow Corning (3140 RTV coating, Midland, MI) was used to encapsulate the wire bonded contact pads on the sensor array. Once the BR-based retinal implant film was placed directly on the foveated imager (Fig. 1, bottom LHS), the number of spikes generated by each pixel was measured under a controlled illumination environment (approximately 3 lux) for a 1-minute duration. The foveated image sensor does not provide colorimetric images, and therefore, a gray scale analysis of the pixels was used to measure the contrast between the two photoproducts of interest. The measured output is equivalent to the pulse frequencies that are proportional to the light intensities monitored by each pixel within the foveated imager.

3. Results and Discussion

3.1. Q state formation

The absorption spectra of the implants were first measured to demonstrate the ability to convert the V49A layers between the two photoproducts (**bR** and **Q** states, Fig. 6). The solid line (1) shows the absorption spectra of the implant in the **bR** state, whereas the dashed line (2) shows the absorption spectra of the implant in the **Q** state. A photograph of the films with the V49A protein in each photostate is shown in the insert. The **Q** state has a transparent/yellow color with an absorption maximum on the periphery of the visible spectrum (~380 nm). This state is reversible, but a high intensity UV light is required to drive the protein back to the photoactive **bR** state, which is characterized by the vibrant, purple color shown in Fig. 6 and has an absorption maximum of 570 nm. Spectral analysis indicates that the film has completely converted from the **bR** resting state to the **Q** state using the red LED light bank (100 mW/cm^2) for 8 hours. There is no spectroscopic evidence of protein denaturation following the immobilization of the protein within the retinal implant architecture, and following the intense red light exposure, the device retained a high optical clarity as suggested by minimal absorption in a wavelength range of > 700 nm.

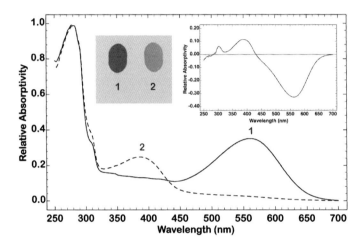

Fig. 6. Absorption spectra of the protein-based retinal implants in the **bR** state (solid line) and the **Q** state (dashed line). The left inset image is a photograph of the retinal implants in the **bR** state (1) and **Q** state (2) after red LED illumination. The right inset figure shows the difference spectra, in which the **bR** state spectrum is subtracted from the **Q** state spectrum.

3.2. *Pixel characterization using a foveated image sensor*

The foveated image sensor was used as a proof-of-concept experiment to demonstrate the capability of monitoring **Q** state formation within a retinal implant architecture. Following a demonstration that the retinal implant could be completely converted to the **Q** photoproduct under intense LED illumination (Fig. 6), the image sensor was subsequently used to quantify differences in pixel intensities through a contrast comparison between the **bR** and **Q** states. Because the foveated imager does not provide a color image, a gray scale was used to measure the difference between the purple pixels of **bR** and the transparent/yellow pixels of the **Q** state. A histogram of the image acquired by the foveated image sensor (Figs. 7(A) and 7(B)), shows a shift in the pixel intensity to the right, reflecting an increase in the number of AP firing events recorded by the pixels (x-axis). The number of active pixels corresponding to the image area that is illuminated is depicted on the y-axis. The observed shift corresponds to the formation of the **Q** state where a brighter image will result in a higher number of APs. In contrast, a "zero" events situation would correspond to the reference background (black color) where no image is detectable, and where only the dark current noise would prevail. This result is expected, since a brighter shade of gray results in a higher light intensity which in turn triggers the photodiodes to generate more APs over a given time domain. Because the **bR** state results in a darker shade of gray, the light intensity falls and the photodiodes would reduce the number of APs generated over the same period. However, a comparable amount of active photodiodes should be present in both the **bR** and **Q** state recording.

Throughout this study, we first recorded the maximum frequency corresponding to complete **Q** state formation, and the lowest frequency, corresponding to the film resting in the **bR** state. Considering Fig. 7, there is an output frequency increase of 2 AP/s (or 120 AP/min) in forming the **Q** state. The inset images for Figs. 7(A) and 7(B) shows the measured pixel intensity distribution map for the retinal implant in both the **bR** and **Q** states, respectively (as the film is placed onto the static pixels of the foveated imager). Note that the curved edges of the retinal implant can be observed in the top right hand corner of each of the inset images, and that the implant was placed in an identical location on the foveated imager between measurements of the **bR** and **Q** states. This placement was performed to ensure that the ambient light conditions and relative pixel measurements were consistent throughout the study. The measured pixel intensities were taken from the same location on the imager for both states, which is shown by the white boxes of the insert images in Fig. 7.

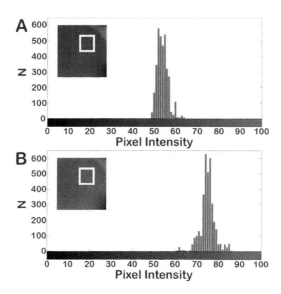

Fig. 7. Quantification of pixel intensity of BR in the **bR** state (A) and the **Q** state (B). The value, N, on the y-axis was measured by the static pixels on the foveated image sensor as the number of events recorded during a 1 minute time period at each pixel intensity. The x-axis provides a graded gray scale used to measure pixel intensities of the retinal implant for each state. The inset images show the retinal implants in gray scale and the corresponding areas of measurement given by the white squares. There is a light intensity variation between each of the photostates of approximately 25%, or 0.85 lux, which translates to 2 AP/s according to Ref. 67.

The pixel intensity analysis demonstrates that there is a significant frequency separation observed between the two accessible photoproducts of the retinal implant based on the V49A BR mutant, and that the protein was uniformly present in each respective photostate within the sensitivity of the imager. This result also mirrors the spectral analysis shown in Fig. 6, in which the protein contained within the retinal implant was shown to completely convert from the **bR** resting state to the **Q** state. While this study does not approach the measurability limits of the foveated image sensor in terms of resolution, sensitivity, and function (i.e., dynamic vs. static pixels), the first steps have been taken towards monitoring two spectrally distinct photoproducts within a multilayered thin film architecture, and have laid the groundwork for future experiments that will further characterize the pixel mediation properties of BR-based retinal implants.

Following these initial proof-of-concept experiments, there are a number of conditions that remain to be optimized for further pixel characterization of the retinal prosthetic. First, the ambient light conditions that were used for measurement should be modulated to promote the greatest output frequency separation between the **bR** and **Q** states. We have so far used ambient laboratory lighting for these measurements, but the use of monochromatic light sources will lead to enhanced performance of the sensor and more controlled contrast and resolution. Once the data collection conditions are optimized to discriminate between frequency bands of firing APs, intermediate frequencies should be measured during the **Q** state formation process in order to monitor incomplete conversion and, perhaps, the kinetics of formation within the retinal implant architecture. Finally, we would like to develop a more sensitive irradiation technique based on fiber optics to locally drive individual pixels or pixel clusters within the retinal implant to the **Q** state. Our global analysis has so far predicted that the foveated imager is capable of differentiating between **bR** and **Q**, however, we hope to further exploit the sensitivity of the CMOS sensor to quantify the limits of pixel mediation within our retinal implant.

4. Conclusion

This work demonstrated the feasibility of using CMOS image sensors, in this case a foveated image sensor, as a microsensor array platform for testing the light sensitivity of two stable photoproducts within a novel photosensitive protein-based retinal implant. The foveated imager measured a significant difference of output frequencies between the **bR** and **Q** states using gray scale analysis. This result also mirrored the spectral properties of the two isolatable photoproducts of BR. Moreover, this investigation represents a first example that demonstrates the possibility of not only fabricating a protein-based retinal implant using a BR mutant, but also to characterize the photoactive behavior of BR that will be responsible for restoring vision at a resolution that is only limited by the density of the protein layers. The **bR** and **Q** states of the high **Q**-forming BR mutant, V49A, provide a photochemical pathway towards pixel mediation within the retinal implant, which will help RP or AMD patients manage the gradual progression of retinal degeneration. Areas of the retina with high retinal degeneration will be targeted with active pixels in the **bR** state, where light-activated proton pumping can be harnessed for retinal stimulation. Deactivation of pixels *via* the accession of **Q** will prevent interference with functional patches of photoreceptor cells. We envision that pixels could be selectively driven to the **Q** state following implantation of the retinal prosthetic using a targeted, multiphoton optical apparatus.

The experiments performed here are the first step in a series of ongoing work to explore the ability to mediate pixels of the BR-based retinal implant, quantify the spatial resolution, and develop the optics necessary for localized **Q** state formation within the retinal implant. In addition to optimizing the experimental conditions for the studies outlined above, we are assembling an optical system for irradiation of areas identical to the size of one pixel of the foveated image sensor to validate the ability of modulating the surface area on a scale relevant to *in vivo* pixel mediation. Finally, we are currently developing a similar microsensor array platform using ion-sensitive field-effect transistors (ISFETs) for the on-chip measurement of the proton-pumping action of BR as a means to quantify spatial sensitivity of the protein-based retinal implant.

Acknowledgment

Work conducted in the laboratory of R.R.B. was supported by grants from the National Institutes of Health (GM-34548, 1R41 EY023461), the National Science Foundation (EMT-0829916, IIP-144822, IIP-1542456), and the Harold S. Schwenk Sr. Distinguished Chair in Chemistry. Researcher L.A.L.F. was funded by the Research Council of Norway under the Leiv Eriksson mobility program (234703/F11). Work in the laboratory of E.A.J. was supported by the Oslofjord Fund (239124).

References

1. D. T. Hartong, E. L. Berson and T. P. Dryja, Retinitis pigmentosa, *Lancet*, **368**, 18-24 (2006).
2. K. M. Gehrs, J. R. Jackson, E. N. Brown, R. Allikmets and G. S. Hageman, Complement, age-related macular degeneration and a vision of the future, *Arch. Ophthalmol.*, **128**, 349-358 (2010).
3. A. Santos, M. S. Humayun, E. de Juan, R. J. Greenburg, M. J. Marsh, I. B. Klock and A. H. Milam, Preservation of the inner retina in retinitis pigmentosa. A morphometric analysis, *Arch. Ophthalmol.*, **115**, 511-515 (1997).
4. J. L. Stone, W. E. Barlow, M. S. Humayun, E. J. de Juan and A. H. Milam, Morphometric analysis of macular photoreceptors and ganglion cells in retinas with retinitis pigmentosa, *Arch. Ophthalmol.*, **110**, 1634-1639 (1992).
5. E. L. Berson, Retinitis pigmentosa: Unfolding its mystery, *Proc. Natl. Acad. Sci. U.S.A.*, **93**, 4526-4528 (1996).
6. R. G. Weleber, Inherited and orphan retinal diseases: Phenotypes, genotypes, and probable treatment groups, *Retina*, **25**, S4-S7 (2005).
7. D. S. Friedman, B. J. O'Colmain, B. Munoz, S. C. Tomany, C. McCarty, P. T. de Jong, B. Nemesure, P. Mitchell and J. Kempen, Prevalence of age-related macular degneration in the United States, *Arch. Ophthalmol.*, **122**, 564-572 (2004).
8. J. P. Hubschman, S. Reddy and S. D. Schwartz, Age-related macular degeneration: Current treatments, *Clin. Ophthalmol.*, **3**, 155-166 (2009).

9. L. A. Wiley, E. R. Burnight, A. E. Songstad, A. V. Drack, R. F. Mullins, E. M. Stone and B. A. Tucker, Patient-specific induced pluripotent stem cells (iPSCs) for the study and treatment of retinal degenerative diseases, *Prog. Retinal Eye Res.*, **44**, 15-35 (2015).
10. P. Y. Baranov, B. A. Tucker and M. J. Young, Low-oxygen culture conditions extend the multipotent properties of human retinal prorgenitor cells, *Tissue Eng., Part A*, **20**, 1465-1475 (2014).
11. W. Chatoo, M. Abdouh, R.-H. Duparc and G. Bernier, Bmi1 distinguishes immature retinal progenitor/stem cells from the main progenitor cell population and is required for normal retinal development, *Stem Cells*, **28**, 1412-1423 (2010).
12. B. P. Hafler, N. Surzenko, K. T. Beier, C. Punzo, J. M. Trimarchi, J. H. Kong and C. L. Cepko, Transcription factor *Olig2* defines subpopulations of retinal progenitor cells biased towards specific cell fates, *Proc. Natl. Acad. Sci. U. S. A.*, **109**, 7882-7887 (2012).
13. J. Luo, P. Y. Baranov, S. Patel, H. Ouyang, J. Quach, F. Wu, A. Qiu, H. Luo, C. Hicks, J. Zeng, J. Zhu, J. Lu, N. Sfeir, C. Wen, M. Zhang, V. Reade, S. Patel, J. Sinden, X. Sun, P. Shaw, M. J. Young and K. Zhang, Human retinal progenitor cell transplantation preserves vision, *J. Biol. Chem.*, **289**, 6362-6371 (2014).
14. S. Schmitt, U. Aftab, C. Jiang, S. Redenti, H. Klassen, E. Miljan, J. Sinden and M. Young, Molecular characterization of human retinal progenitor cells, *Invest. Ophthalmol. Vis. Sci.*, **50**, 5901-5908 (2009).
15. A. Gonzalez-Cordero, E. L. West, R. A. Pearson, Y. Duran, L. S. Carvalho, C. J. Chu, A. Naeem, S. J. Blackford, A. Georgiadis, J. Lakowski, M. Hubank, A. J. Smith, J. W. Bainbridge, J. C. Sowden and R. R. Ali, Photoreceptor precursors derived from three-dimensional embryonic stem cell cultures integrate and mature within an adult degenerate retina, *Nat. Biotechnol.*, **31**, 741-747 (2013).
16. S. D. Schwartz, J. P. Hubschman, G. Heilwell, V. Franco-Cardenas, C. K. Pan, R. M. Ostrick, E. Mickunas, R. Gay, I. Klimanskaya and R. Lanza, Embryonic stem cell trials for macular degeneration: A preliminary report, *Lancet*, **379**, 713-720 (2012).
17. A. O. Cramer and R. E. MacLaren, Clinical trials using induced pluripotent stem cells from bench to bedside: Applicaiton to retinal diseases, *Curr. Gene Ther.*, **13**, 139-151 (2013).
18. L. Buchen, Neuroscience: Illuminating the brain, *Nature*, **465**, 26-28 (2010).
19. E. S. Boyden, F. Zhang, E. Bamberg, G. Nagel and K. Deisseroth, Millisecond-timescale, genetically targeted optical control of neural activity, *Nat. Neurosci.*, **8**, 1263-1268 (2005).
20. X. Han and E. S. Boyden, Multiple-color optical activation, silencing, and desynchronization of neural activity, with single-spike temporal resolution, *PLoS One*, **2**, e299 (2007).
21. V. Busskamp and B. Roska, Optogenetic approaches to restoring visual function in retinities pigmentosa, *Curr. Opin. Neurobiol.*, **21**, 942-946 (2011).
22. V. Busskamp, S. Picaud, J. A. Sahel and B. Roska, Optogenetic therapy for retinitis pigmentosa, *Gene Ther.*, **19**, 1-7 (2011).
23. P. Degenaar, N. Grossman, M. A. Memon, J. Burrone, M. Dawson, E. Drakakis, M. Neil and K. Nikolic, Optobionic vision – a new genetically enhanced light on retinal prosthesis, *J. Neural. Eng.*, **6** (2009).
24. J. Cehajic-Kapetanovic, C. Eleftheriou, A. E. Allen, N. Milosavljevic, A. Pienaar, R. Bedford, K. E. Davis, P. N. Bishop and R. J. Lucas, Restoration of vision with ectopic expression of human rod opsin, *Curr. Biol.*, **25**, 2111-2122 (2015).
25. K. Shintani, D. L. Shechtman and A. S. Gurwood, Review and update: Current treatment trends for patients with retinitis pigmentosa, *J. Am. Optom. Assoc.*, **80**, 384-401 (2009).
26. J. L. Kovach, S. G. Schwartz, H. W. Flynn Jr. and I. U. Scott, Anti-VEGF treatment strategies for wet AMD, *J. Opthalmol.*, **2012** (2012).
27. W. M. Grill and J. T. Mortimer, Stimulus waveforms for selective neural stimulation, *IEEE Eng. Med. Biol. Mag.*, **14**, 375-385 (1995).

28. J. D. Weiland, A. K. Cho and M. S. Humayun, Retinal prostheses: Current clinical results and future needs, *Ophthalmology*, **118**, 2227-2237 (2011).
29. M. S. Humayun, J. D. Dorn, L. da Cruz, G. Dagnelie, J.-A. Sahel, P. E. Stanga, A. V. Cideciyan, J. L. Duncan, D. Eliott, E. Filley, A. C. Ho, A. Santos, A. B. Safran, A. Arditi, L. B. Del Priore and R. J. Greenberg, Interim results from the international trial of Second Sight's visual prosthesis, *Ophthalmology*, **119**, 779-788 (2012).
30. M. P. Barry and G. Dagnelie, Use of the Argus II retinal prosthesis to improve visual guidance of fine hand movements, *Invest. Ophthalmol. Vis. Sci.*, **53**, 5095-5101 (2012).
31. R. Wilke, V.-P. Gabel, H. Sachs, K.-U. Bartz-Schmidt, F. Gekeler, D. Besch, P. Szurman, A. Stett, B. Wilhelm, T. Peters, A. Harscher, U. Greppmaier, S. Kibbel, H. Benhav, A. Bruckmann, K. Stingl, A. Kusnyerik and E. Zrenner, Spatial resolution and perception of patterns mediated by a subretinal 16-electrode array in patients blinded by hereditary retinal dystrophies, *Invest. Ophthalmol. Vis. Sci.*, **52**, 5995-6003 (2011).
32. H. Benav, K. U. Bartz Schmidt, D. Besch, A. Bruckmann, F. Gekeler, U. Greppmaier, A. Harscher, S. Kibbel, A. Kusnyerik, T. Peters, H. Sachs, A. Stett, K. Stingl, B. Wilhelm, R. Wilke, W. Wrobel and E. Zrenner, Restoration of useful vision up to letter recognition capabilities using subretinal microphotodiodes, *Conf. Proc. IEEE Eng. Med. Biol. Soc.*, **2010**, 5919-5922 (2010).
33. A. Kusnyerik, U. Greppmaier, R. Wilke, F. Gekeler, B. Wilhelm, H. Sachs, K. U. Bartz-Schmidt, U. Klose, K. Stingl, M. D. Resch, A. Hekmat, A. Bruckmann, K. Karacs, J. Nemeth, I. Suveges and E. Zrenner, Positioning of electronic subretinal implants in blind retinitis pigmentosa patients through multimodal assessment of retinal structures, *Invest. Ophthalmol. Vis. Sci.*, **53**, 3748-3755 (2012).
34. K. Stingl, K. U. Bartz-Schmidt, D. Besch, A. Braun, A. Bruckmann, F. Gekeler, U. Greppmaier, S. Hipp, G. Hortdorfer, C. Kernstock, A. Koitschev, A. Kusnyerik, H. G. Sachs, A. Schatz, K. T. Stingl, T. Peters, B. Wilhelm and E. Zrenner, Artificial vision with wirelessly powered subretinal electronic implant alpha-IMS, *Proc. R. Soc. B*, **280**, 20130077 (2013).
35. E. Zrenner, Fighting blindness with microelectronics, *Sci. Transl. Med.*, **5**, 210-216 (2013).
36. S. Klauke, M. Goertz, S. Rein, D. Hoehl, U. Thomas, R. Eckhorn, F. Bremmer and T. Wachtler, Stimulation with a wireless intraocular epiretinal implant elicits visual percepts in blind humans, *Invest. Ophthalmol. Vis. Sci.*, **52**, 449-455 (2011).
37. G. Roessler, T. Laube, C. Brockmann, T. Kirschkamp, B. Mazinani, M. Goertz, C. Koch, I. Krisch, B. Sellhaus, H. Khiem Trieu, J. Weis, N. Bornfeld, H. Röthgen, A. Messner, W. Mokwa and P. Walter, Implantation and explanation of a wireless epiretinal retina implant device: Observations during the EPIRET3 prospective clinical trial, *Invest. Ophthalmol. Vis. Sci.*, **50**, 3003-3008 (2009).
38. J. Menzel-Severing, T. Laube, C. Brockmann, N. Bornfield, W. Mokwa, B. Mazinani, P. Walter and G. Roessler, Implantation and explantation of an active epiretinal visual prosthesis: 2-year follow-up data from the EPIRET3 prospective clinical trial, *Eye*, **26**, 501-509 (2012).
39. M. Javaheri, D. S. Hahn, R. R. Lakhanpal, J. D. Weiland and M. S. Humayun, Retinal prostheses for the blind, *Ann. Acad. Med. Singapore*, **35**, 137-144 (2006).
40. D. Palanker, A. Vankov, P. Huie and S. Baccus, Design of a high-resolution optoelectronic retinal prosthesis, *J. Neural. Eng.*, **2**, S105-S120 (2005).
41. E. Zrenner, Fighting blindness with microelectronics, *Sci. Transl. Med.*, **5**, 1-7 (2013).
42. J. D. Dorn, A. K. Ahuja, A. Caspi, L. da Cruz, G. Dagnelie, J. -A. Sahel, R. J. Greenberg and M. J. McMahon, The detection of motion by blind subjects with the epiretinal 60-electrode (Argus II) retinal prosthesis, *JAMA Ophthalmol.*, **131**, 183-189 (2013).
43. T. T. Kien, T. Maul and A. Bargiela, A review of retinal prosthesis approaches, *Int. J. Mod. Phys.: Conf. Ser.*, **9**, 209-231 (2012).

44. N. L. Wagner, J. A. Greco and R. R. Birge, Visual restoration using microbial rhodopsin, In *Bionanotechnology: Biological Self-Assembly and its Applications*, B. H. A. Rehm, Ed., Caister Academic Press, Norfolk, UK, pp. 205-240 (2013).
45. J.-A. He, L. Samuelson, L. Li, J. Kumar and S. K. Tripathy, Oriented bacteriorhodopsin/ polycation multilayers by electrostatic layer-by-layer assembly, *Langmuir*, **14**, 1674-1679 (1998).
46. N. L. Wagner, J. A. Greco, M. J. Ranaghan and R. R. Birge, Directed evolution of bacteriorhodopsin for applications in bioelectronics, *J. R. Soc., Interface*, **10**, 20130197 (2013).
47. D. Oesterhelt and W. Stoeckenius, Functions of a new photoreceptor membrane, *Proc. Natl. Acad. Sci. U.S.A.*, **70**, 2853-2857 (1973).
48. D. Oesterhelt and W. Stoeckenius, Rhodopsin-like protein from the purple membrane of Halobacterium halobium, *Nature (London), New Biol.*, **233**, 149-152 (1971).
49. H. Luecke, B. Schobert, H.-T. Richter, J.-P. Cartailler and J. K. Lanyi, Structure of bacteriorhodopsin at 1.55 Å resolution, *J. Mol. Biol.*, **291**, 899-911 (1999).
50. C. D. Heyes and M. A. El-Sayed, The role of the native lipids and lattice structure in bacteriorhodopsin protein conformation and stability as studied by temperature-dependent fourier transform-infrared spectroscopy, *J. Biol. Chem.*, **277**, 29437-29443 (2002).
51. R. Bogomolni, R. Baker, R. Lozier and W. Stoeckenius, Light-driven proton translocations in Halobacterium halobium, *Biochim. Biophys. Acta, Bioenerg.*, **440**, 68-88 (1976).
52. R. R. Birge, N. B. Gillespie, E. W. Izaguirre, A. Kusnetzow, A. F. Lawrence, D. Singh, Q. W. Song, E. Schmidt, J. A. Stuart, S. Seetharaman and K. J. Wise, Biomolecular electronics: Protein-based associative processors and volumetric memories, *J. Phys. Chem. B*, **103**, 10746-10766 (1999).
53. J. Tittor and D. Oesterhelt, The quantum yield of bacteriorhodopsin, *FEBS Lett.*, **263**, 269-273 (1990).
54. R. R. Birge, T. M. Cooper, A. F. Lawrence, M. B. Masthay, C. Vasilakis, C. F. Zhang and R. Zidovetzki, A spectroscopic, photocalorimetric and theoretical investigation of the quantum efficiency of the primary event in bacteriorhodopsin, *J. Am. Chem. Soc.*, **111**, 4063-4074 (1989).
55. M. B. Jackson and J. M. Sturtevant, Phase transitions of the purple membranes of Halobacterium halobium, *Biochemistry*, **17**, 911-915 (1978).
56. R. R. Birge, Photophysics and molecular electronic applications of the rhodopsins, *Annu. Rev. Phys. Chem.*, **41**, 683-733 (1990).
57. C. Bräuchle, N. Hampp and D. Oesterhelt, Optical applications of bacteriorhodopsin and its mutated variants, *Adv. Mater.*, **3**, 420-428 (1991).
58. M. J. Ranaghan, S. Shima, L. Ramos, D. S. Poulin, G. Whited, S. Rajasekaran, J. A. Stuart, A. D. Albert and R. R. Birge, Photochemical and thermal stability of green and blue proteorhodopsins: Implications for protein-based bioelectronic devices, *J. Phys. Chem. B*, **114**, 14064-14070 (2010). N. Hampp, Bacteriorhodopsin as a photochromic retinal protein for optical memories, *Chem. Rev.*, **100**, 1755-1776 (2000).
59. N. Hampp, Bacteriorhodopsin as a photochromic retinal protein for optical memories, *Chem. Rev.*, **100**, 1755-1776 (2000).
60. A. Popp, M. Wolperdinger, N. Hampp, C. Bräuchle and D. Oesterhelt, Photochemical conversion of the O-intermediate to 9-cis-retinal-containing products in bacteriorhodopsin films, *Biophys. J.*, **65**, 1449-1459 (1993).
61. N. B. Gillespie, K. J. Wise, L. Ren, J. A. Stuart, D. L. Marcy, J. Hillebrecht, Q. Li, L. Ramos, K. Jordan, S. Fyvie and R. R. Birge, Characterization of the branched-photocycle intermediates P and Q of bacteriorhodopsin, *J. Phys. Chem. B*, **106**, 13352-13361 (2002).
62. Z. Dáncshazy and Z. Tokaji, Blue light regeneration of bacteriorhodopsin bleached by continuous light, *FEBS Lett.*, **476**, 171-173 (2000).

63. M. J. Ranaghan, J. A. Greco, N. L. Wagner, R. Grewal, R. Rangarajan, J. F. Koscielecki, K. J. Wise and R. R. Birge, Photochromic bacteriorhodopsin mutant with high holographic efficiency and enhanced stability via a putative self-repair mechanism, *ACS Appl. Mater. Interfaces*, **6**, 2799-2808 (2014).
64. J. A. Stuart, D. L. Marcy, K. J. Wise and R. R. Birge, Volumetric optical memory based on bacteriorhodopsin, *Synth. Met.*, **127**, 3-15 (2002).
65. J. A. Stuart, J. R. Tallent, E. H. L. Tan and R. R. Birge, Protein-based volumetric memories, *Proc. IEEE Nonvol. Mem. Tech. (INVMTC)*, **6**, 45-51 (1996).
66. J. A. Greco, N. L. Wagner, M. J. Ranaghan, S. Rajasekaran and R. R. Birge, Protein-based optical computing, In *Biomolecular Information Processing: From Logic Systems to Smart Sensors and Actuators*, E. Katz, Ed., Wiley-VCH: Weinheim, Germany, pp. 33-59 (2012).
67. M. Azadmehr, J. P. Abrahamsen and P. Häfliger, *IEEE International Symposium on Circuits and Systems (ISCAS)*, **3**, pp. 2751-2754 (2005).
68. R. F. Peck, S. DasSarma and M. P. Krebs, Homologous gene knockout in the archaeon *Halobacterium salinarum* with ura3 as a counterselectable marker, *Mol. Microbiol.*, **35**, 667-676 (2000).
69. B. M. Becher and J. Y. Cassim, Improved isolation procedures for the purple membrane of halobacterium halobium, *Prep. Biochem.*, **5**, 161-178 (1975).
70. D. Oesterhelt and W. Stoeckenius, Isolation of the cell membrane of *Halobacterium halobium* and its fractionation into red and purple membrane, *Methods Enzymol.*, **31**, 667-678 (1974).
71. J.-A. He, L. Samuelson, L. Li, J. Kumar and S. K. Tripathy, Photoelectric properties of oriented bacteriorhodopsin / polycation multilayers by electrostatic layer-by-layer assembly, *J. Phys. Chem. B*, **102**, 7067-7072 (1998).
72. J.-A. He, L. Samuelson, L. Li, J. Kumar and S. K. Tripathy, Bacteriorhodopsin thin film assemblies – immobilization, properties, and applications, *Adv. Mater.*, **11**, 435-446 (1999).
73. C. Scholz, Perspectives on: Materials aspects for retinal prostheses, *J. Bioact. Compat. Polym.*, **22**, 539-568 (2007).
74. S. Roll, J. Müller-Nordhorn, T. Keil, H. Scholz, D. Eidt, W. Greiner and S. N. Willich, Dacron® vs. PTFE as bypass materials in peripheral vascular surgery – systematic review and meta-analysis, *BMC Surg.*, **8**, 1-8 (2008).
75. Y. Liu, T. He, H. Song and C. Gao, Layer-by-layer assembly of biomacromolecules on poly(ethylene terephthalate) films and fiber fabrics to promote endothelial cell growth, *J. Biomed. Mat. Res*, Part A, **81A**, 692-704 (2007).
76. Y. Liu, T. He and C. Gao, Surface modification of poly(ethylene terephthalate) via hydroysis and layer-by-layer assembly of chitosan and chondroitin sulfate to construct cytocompatible layer for human endothelial cells, *Coll. Surf., B*, **46**, 117-126 (2005).
77. M. C. Wyers, M. D. Phaneuf, E. M. Rzucidlo, M. A. Contreras, F. W. LoGerfo and W. C. Quist, In vivo assessment of a novel dacron surface with covalently bound recombinant hirudin, *Cardiovasc. Pathol.*, **8**, 153-159 (1999).
78. A. Matsuzawa, Nano-scale CMOS and low voltage analog to digital converter design challenges, *8th International Conference on Solid-State and Integrated Circuit Technology*, pp. 1676-1679 (2006).
79. A. El Gamal, Trends in CMOS image sensor technology and design, *International Electron Devices Meeting*, pp. 805-808 (2002).
80. C. T. Chiang and C. Y. Wu, Implantable neuromorphic vision chips, *Electron. Lett.*, **40**, 361-363 (2004).
81. L. Wentai, K. Vichienchom, M. Clements, S. C. DeMarco, C. Hughes, E. McGucken, M. S. Humayun, E. De Juan, J. D. Weiland and R. J. Greenberg, A neuro-stimulus chip with telemetry unit for retinal prosthetic device, *IEEE J. Solid-State Circuits*, **35**, 1487-1497 (2000).

82. T. Delbrück and C. A. Mead, An electronic photoreceptor sensitive to small changes in intensity, In *Advances in Neural Information Processing Systems 1*, S. T. David, Ed.; Morgan Kaufmann Publishers Inc., pp. 720-727 (1989).
83. C. Mead, *Analog VLSI and Neural Systems* (Addison-Wesley, 1989).
84. A. Mortara and E. A. Vittoz, A communication architecture tailored for analog vlsi artificial neural networks: Intrinsic performance and limitations, *IEEE Trans. Neural Networks*, **5**, 459-466 (1994).

A Low-Power Low-Data Rate Impulse Radio Ultra-Wideband (IR-UWB) Transmitter

Ifana Mahbub[*], Samira Shamsir and Syed K. Islam

Department of Electrical Engineering and Computer Science,
University of Tennessee, Knoxville, TN 37996, USA
[*]imahbub@vols.utk.edu

A low-power and low-data-rate (100 kbps) fully integrated CMOS impulse radio ultra-wideband (IR-UWB) transmitter for biomedical application is presented in this paper. The transmitter is designed using a standard 180-nm CMOS technology that operates at the 3.1-5 GHz frequency range with more than 500 MHz of channel bandwidth. Modulation scheme of this transmitter is based on on-off keying (OOK) in which a short pulse represents binary "1" and absence of a pulse represents binary "0" transmission. During the 'off' state (sleep mode) the transmitter consumes only 0.4 µW of power for an operating voltage of 1.8 V while during the impulse transmission state it consumes a power of 36.29 µW. A pulse duration of about 3.5 ns and a peak amplitude of the frequency spectrum of about -47.8 dBm/MHz are obtained in the simulation result which fully complies with Federal Communication Commission (FCC) regulation.

Keywords: ultra-wideband (UWB); low-power; transmitter; CMOS.

1. Introduction

It is of crucial importance to design a transmitter that operates in the desired allocated spectrum and does not interfere with the co-existing bands since the interference from other bands can significantly deteriorate the radio performance. To minimize this problem as much as possible Federal Communications Commission (FCC) has limited the radiation characteristics of a radio operating in certain bands. Unlike GSM or CDMA which work in a licensed spectrum, FCC has also approved many unlicensed bands where the radio only needs to comply with the emission mask. Since 2002 when FCC allocated 3.1 to 10.6 GHz of the frequency spectrum as *ultra-wideband*, many transmitter designs have been reported with a pulse generator and oscillator to up-convert the baseband signal to RF signal [1], [2]. For a certain commercial biomedical application, the cost is also an important factor, which can be minimized by making the wireless telemetry operate in the unlicensed ultra-wideband (UWB). In this paper, the UWB emission mask imposed by the FCC and the design of the impulse radio ultra-wideband (IR-UWB) transmitter is presented that follows the spectral mask constraints.

[*]Corresponding author.

2. UWB Communication and FCC Spectral Mask

Pulsed based communication requires a broad bandwidth as the narrow impulse signal in the time domain corresponds to a wider bandwidth in the frequency domain. That is why this type of communication is also known as ultra-wideband (UWB) communication. Traditional continuous wave (CW) signal is referred to as a narrowband signal as the frequency bandwidth is very low. According to the regulation of the Federal Communications Commission (FCC), the bandwidth of a UWB signal must be greater than 500 MHz and the transmitted power density must be less than -41.3 dBm/MHz over the 3.1-10.6 GHz frequency band [3]. The fractional bandwidth is defined as, $(f_h - f_l)/f_c$ where f_h and f_l are the upper and the lower -10 dB cut-off frequencies and f_c is the center frequency. Because of the narrow pulse nature of the UWB communication, it is possible to accurately localize a subject (for example senior citizen for assisted living) [4] or even monitor the rhythmic chest movement due to respiration [5]. It has also been extensively utilized for high data rate communication applications. Chae et al. has demonstrated a neural recording system for neuron spike extraction with 128 recording channel and 90 Mbps data rate [6]. All of these unique features of UWB communication offer many attractive areas of research.

In February 2002, FCC released a regulation for the UWB communication power spectrum emission. Figure 1 shows the FCC spectral mask for UWB communication. According to the FCC regulation, the maximum emission power must not exceed -41.3 dBm/MHz so that it does not interfere with other medical devices operating in the Wi-fi or Bluetooth frequency bands because of its low transmission power. FCC has also imposed an even higher restriction on the emission mask for outdoor communication compared to indoor communication so that there is no interference with cellular or aircraft radar communication. A deep notch around 1.5 GHz is imposed to avoid interference with the Global Positioning System (GPS). Thus the UWB communication does not impose any

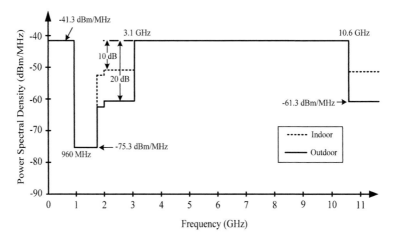

Fig. 1. FCC mask for UWB communication.

interference with the existing bands such as 2.4 GHz ISM band. Because of the lower emission power, the range of communication is also limited to less than 10 m distance.

3. UWB Pulse Shaping Techniques

Many approaches have been studied over the last one and a half decade to implement the pulse shaping technique to achieve the desired UWB spectrum that follows the FCC mask regulation. These approaches can be broadly categorized as the carrier-based and carrier-less UWB communication. For a carrier-less UWB radio, the pulse spectrum falls within the 0-960 MHz or 3.1 -10.6 GHz bandwidth, without the requirement of RF up-conversion. The duration of the pulse is less than 1 ns with one or more Gaussian function like wavelet. Impulse generators have been implemented that generates Gaussian waveform or its derivative for carrier-less UWB communication. A filter is sometimes incorporated after the pulse generator to make sure that the power spectrum lies within the FCC spectral mask [7]. The higher order derivative of the Gaussian pulse has a sharper roll-off which helps in meeting FCC spectrum regulation without using any additional filter. Kim *et al.* presented a 5th derivative Gaussian impulse waveform generator using digital CMOS circuitry that does not require any additional filtering and occupies 7.2 GHz of bandwidth [8]. It is difficult to calibrate the variations of standard deviation parameter and the derivative order of the Gaussian wavelet due to process and temperature variation (PVT), violating the FCC spectral mask regulation. In the carrier-based UWB communication system, the UWB spectrum is sub-divided into several bands, each with a bandwidth of 500 MHz or higher, as shown in Fig. 2. Carrier-based UWB communication enables multi-user application, for example, where each user would be assigned particular center frequency and bandwidth in the multi-band UWB system. In the sub-banded UWB system, the baseband data is first converted to a particular pulse shape, (e.g. triangular, sinc, square, etc.) with narrow width and then the wavelet is up-converted by a local oscillator (LO) with the help of a mixer, as shown in Fig. 3.

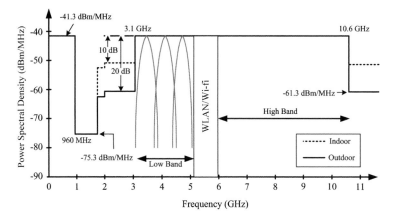

Fig. 2. Carrier-based UWB communication system.

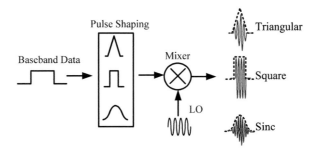

Fig. 3. Carrier-based UWB wavelet generation scheme.

For choosing the right pulse shaping that would be optimum for any particular application designers first have to look into the spectral efficiency, out-of-band emission and side-lobe rejection of each shape. Spectral efficiency or bandwidth efficiency is a measure of how efficiently the spectrum of the pulse wavelet occupies the given bandwidth. If the data rate is R and the bandwidth is B, then the spectral efficiency is defined as R/B (bits per second per Hz). The sinc shaped pulse has a spectral efficiency of 100% because of having rectangular spectrum in frequency domain while the square-shaped pulse has a spectral efficiency of only 58% due to having sinc shaped spectrum [9]. Out-of-band emission is also an important factor while choosing the optimum pulse shaping which represents the power emitted by the transmitter in the adjacent channel which is susceptible to interference with other radio communication in the nearby frequency bands. The sinc wave shape has 0% out-of-band emission while the square wave has 12.8% (-8.9 dB) out-of-band emission [9]. Side-lobe rejection primarily displays the severity of the out-of-band emission. It is the difference of the power emitted in-band and out-of-band. The sinc and Gaussian waveforms have no side-lobes and the side-lobe rejection for square-shaped wavelet is -13 dB [9]. In the present work, both triangular pulse shaping and sinc pulse shaping are implemented.

4. Impulse Radio UWB Transmitter for Low-Data-Rate Application

For any biomedical sensor application, the transmitter is expected to transmit data wirelessly at a steady low data rate. Since the transmitter usually consumes most of the power of a sensor system, any reduction in the power consumption of the transmitter block would save the overall power consumption of the system. To reduce the average power consumption, it makes sense to duty-cycle the radio communication as much as possible so that the transmitter is "ON" for a short period and thus dissipating power only when it is active. Figure 4(a) shows the duty-cycling approach, where the transmitter is "OFF" most of the time, thus dissipating only the leakage current. Unlike conventional continuous wave (CW) transmitter where data is encoded by the phase or the frequency of the carrier waveform, the data can also be encoded using short duration pulses. This type of radio communication is known as Impulse Radio Ultra-Wideband (IR-UWB) because of their pulsed nature. It is the transmission of very low energy narrow pulses (in the range of sub-nanoseconds to few nanoseconds) with fine time resolution. Although IR-UWB has been a popular choice for high data rate application for its wide bandwidth nature, it has potential to be implement in low-data-rate biomedical applications as well considering the lower power consumption. For the CW transmitters or the duty-cycled IR-UWB transmitter, the static power (leakage and overhead power) consumption is fixed no matter what the data rate is. It is the dynamic power that increases with the increasing data rate. The dynamic power can be expressed as the multiplication of E_b and R, where E_b is the energy per bit and R is the data rate. For an IR-UWB transmitter E_b is much lower than a CW transmitter, thus resulting in a reduced overall power consumption of the transmitter (Fig. 4(b)). For the receivers in wearable biomedical sensor applications most of the time there is not much of a power restriction, thus these receivers can always be active for communication. For IR-UWB communication many modulation schemes such OOK (On-off-keying) [1], Pulse Amplitude Modulation (PAM) [10], Binary Phase Shift Keying (BPSK) [11], Pulse Position Modulation (PPM) [12] have been investigated. Figure 5 shows various modulation schemes. With the motivation of achieving simplicity in transmitter architecture and low-power design, OOK modulation is an ideal candidate for low-data-rate application. In healthcare applications in a hospital environment, the patient data is collected from multiple sensors. Thus the carrier based IR-UWB architecture working in 3.1 to 5 GHz frequency range with OOK modulation scheme is chosen for the transmitter design which can be integrated with sensors and signal processing electronics in patient monitoring systems. CMOS process has been selected for the design of the transmitter considering its availability and low-cost fabrication. The following section describes the transmitter architecture that was designed and fabricated using 180nm high voltage CMOS process.

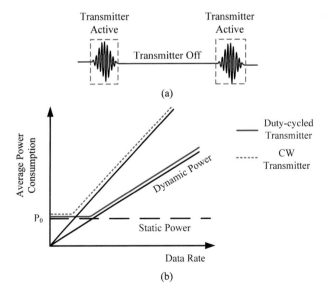

Fig. 4. (a) Transmitter duty-cycling approach and (b) Static and dynamic power consumption of duty-cycled and CW transmitter.

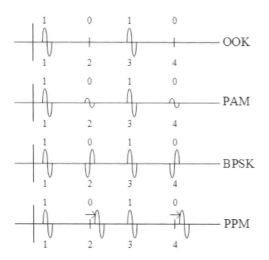

Fig. 5. Different modulation schemes in IR-UWB communication system.

5. Design of IR-UWB Transmitter in 180nm CMOS Process

The system level design of the proposed IR-UWB transmitter is illustrated schematically in Figs. 6-10. In the transmitter end, any additional up-conversion mixer is not required which makes the impulse-based UWB communication a carrier-less radio technology. Its power consumption is much lower than the conventional continuous wave (CW) radio-frequency (RF) systems because in its modulation scheme the generated pulses are passed directly to the antenna or through a buffer. The block diagram of the transmitter consists of an impulse generator, an inverter chain buffer, a tunable cross-coupled LC oscillator, and a power amplifier. In order to reduce the average power consumption, the buffer is enabled only during the pulse emission and disabled during the absence of a pulse transmission as it consumes the majority of the current during the sleep mode. The core building block of the impulse generator is shown in Fig. 7 which includes a tunable delay cell, an inverter, a NAND gate and finally another inverter to generate the desired impulse signal. The biasing voltage $V1$ of the current-starved inverters as shown in Fig. 8 controls the delay in the delay cell block. The inverter chain buffer is used to drive the bond-wire inductance and the parasitic capacitance due to the QFN packaging and the bond pads. The LC voltage controlled oscillator (VCO) consists of a cross-coupled NMOS pair (M_{2-3}) and LC tank with inductors (L_{1-2}) and variable capacitors (C_{var}) as shown in Fig. 9. The control voltage, V_{CTRL} can tune the center frequency of oscillation and the frequency can again be fine-tuned by the capacitor banks. The switches that are connected to the capacitors in the bank are turned on/off by using digital signals V_{0-1} while M_1 is turned on/off by using impulse signal coming from the impulse generator block. The LC VCO block is isolated from the load impedance variation by a cascaded buffer. The LC VCO block turns on only during the pulse generation and turns off for the rest of the time to achieve higher energy efficiency. In order to drive a 50 Ω antenna, the source-follower topology is chosen for the

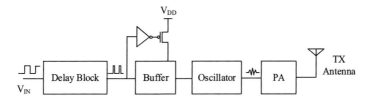

Fig. 6. System level block diagram.

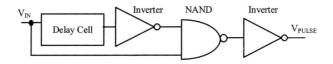

Fig. 7. Impulse generator block.

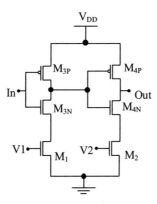

Fig. 8. Delay cell circuit.

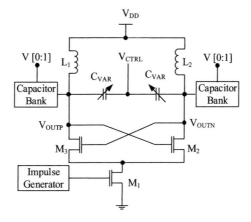

Fig. 9. Tunable LC oscillator circuit.

buffer stage with proper sizing and bias [13]. The power amplifier is designed using cascade topology with resistive shunt feedback as shown in Fig. 10. An output waveform of a 3.5 GHz UWB pulse obtained by an input signal (V_{in}) of 100 kbps data rate is shown in Fig. 11. The differential output signal from the LC VCO is converted to a single-ended output by using a 50 Ω balun. As shown in Fig. 12, the simulated pulse width is obtained as 3.5 ns and the peak-to-peak output voltage amplitude is 1.5 V. From Fig. 13, it can be seen that 3.86 GHz is the center of the spectrum of the impulse signal. According to the convention, the bandwidth is measured at 10 dB below the peak power level of -47.8 dBm/MHz and obtained to be 2 GHz which compiles with the FCC mask.

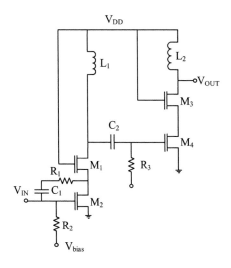

Fig. 10. Circuit schematic of the PA.

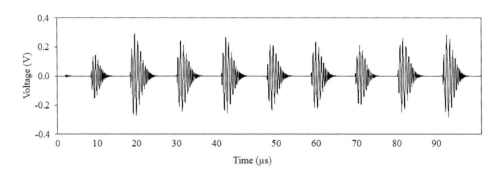

Fig. 11. Impulse waveform output at 100kbps data rate.

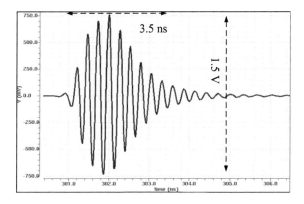

Fig. 12. Simulation result of the impulse waveform (zoomed) from the LC oscillator.

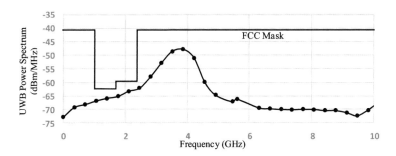

Fig. 13. FFT of the oscillator output signal.

6. Conclusion

This paper presents an OOK based IR-UWB transmitter designed using a standard 180 nm CMOS process. OOK based modulation based transmitter architecture is adopted to achieve a low-data-rate (100 kbps) in motivation to be use in biomedical applications. Having low power consumption by heavy duty-cycling and low cost by using unlicensed ultra-wideband, IR-UWB transmitter can be a potential candidate in low power wireless communication in biomedical sensor application. Circuit level design along with the simulation results of such transmitter are presented. The power consumption for this design is much lower than the conventional CW-RF system due to the implementation of carrier-less impulse-based UWB technology that eliminates additional up-conversion mixer in the transmitter end. In addition, the obtained bandwidth and peak power level are fully compatible with the FCC regulations.

References

1. T. A. Phan, J. Lee, V. Krizhanovskii, S. K. Han, and S. G. Lee, "A 18-pJ/Pulse OOK CMOS Transmitter for Multiband UWB Impulse Radio", *IEEE Microwave and Wireless Components Lett.* **17** (2007) 688-690.

2. D. D. Wentzloff and A. P. Chandrakasan, "Gaussian pulse Generators for subbanded ultra-wideband transmitters", *IEEE Trans. on Microwave Theory and Techniques.* **54** (2006) 1647-1655.
3. "Revision of Part 15 the Commission's rules regarding ultra-wideband transmission systems", *FCC.* **ET Docket** (2002) 98–153.
4. M. A. Stelios, A. D. Nick, M. T. Effie, K. M. Dimitris, and S. C. Thomopoulos, "An indoor localization platform for ambient assisted living using UWB", in *Proc. 6th Int. Conf. on Adv. in Mobile Computing and Multimedia*, 2008, 178-182.
5. K. Higashikaturagi, Y. Nakahata, I. Matsunami, and A. Kajiwara, "Non-invasive respiration monitoring sensor using UWB-IR", *IEEE Int. Conf. on Ultra-Wideband*, 2008, 101-104.
6. M. S. Chae, Z. Yang, M. R. Yuce, L. Hoang, and W. Liu, "A 128-Channel 6 mW Wireless Neural Recording IC with Spike Feature Extraction and UWB Transmitter", *IEEE Trans. on Neural Sys. and Rehabilitation Eng.* **17** (2009) 312-321.
7. Y. Jeong, S. Jung, and J. Liu, "A CMOS impulse generator for UWB wireless communication systems", in *Circuits and Systems, 2004. ISCAS'04. Proc. of the 2004 Int. Symposium on.* **4** (2004) IV-129-32.
8. H. Kim, D. Park, and Y. Joo, "All-digital low-power CMOS pulse generator for UWB system", *Elec. Lett.,* **40** (2004).
9. A. Apsel, X. Wang, and R. Dokania, "Low Power Impulse Radio Transceivers", *Design of Ultra-Low Power Impulse Radios*, Springer (2014) 37-69.
10. S. Hongsan, P. Orlik, A. M. Haimovich, L. J. Cimini, and Z. Jinyun, "On the spectral and power requirements for ultra-wideband transmission", in *Communications, 2003. ICC '03. IEEE International Conference on*, 2003, pp. 738-742.
11. M. Demirkan and R. R. Spencer, "A Pulse-Based Ultra-Wideband Transmitter in 90-nm CMOS for WPANs", *IEEE Journal of Solid-State Circuits,* **43** (2008) 2820-2828.
12. F. Padovan, A. Bevilacqua, and A. Neviani, "A 20Mb/s, 2.76 pJ/b UWB impulse radio TX with 11.7% efficiency in 130 nm CMOS", *European Solid State Circuits Conference (ESSCIRC),* 2014, pp. 287-290.
13. J. Lee, Y. Park, M. Kim, C. Yoon, J. Kim, and K. Kim, "System-on package ultra-wideband transmitter using CMOS impulse generator", *IEEE Trans. Microwave Theory Tech.*, **54** (4) (2006) 1667-1674.

Multi-Bit NVRAMs Using Quantum Dot Gate Access Channel

Murali Lingalugari, Pik-Yiu Chan, John Chandy, Evan Heller[†] and Faquir Jain[*]

Electrical and Computer Engineering,
University of Connecticut,
371 Fairfield Way, Unit 4157, Storrs, CT 06269, USA
[*]*fcj@engr.uconn.edu*
[†]*Synopsis Inc., Ossining, NY 10562, USA*
evankheller@gmail.com

This paper presents a quantum dot access channel nonvolatile random access memory (QDAC-NVRAM) which has comparable write and erase times to conventional random access memories but consumes less power and has a smaller footprint. We have fabricated long-channel (W/L = 15μm/10μm) nonvolatile random access memories (NVRAMs) with 4μs erase times. These devices are CMOS-compatible and employ novel quantum dot access channel (QDAC) which enables fast storage and retrieval of charge from the floating gate layer. In addition, QDNVRAMs are shown to be capable of storing multiple-bits and potentially scalable to sub 22nm. We are also presenting the simulation results. This paper also presents a memory array architecture using QDAC-NVRAMs.

Keywords: cladded quantum dots; site-specific self-assembly; quantum dot floating gate nonvolatile memory; nonvolatile random access memory; quantum dot channel FET; quantum dot access channel, low-voltage and high-speed erase; multi-bit storage capability.

1. Introduction

Emerging nonvolatile random access memory (NVRAM) technologies include phase change memories (PCMs), magnetic random access memories (MRAMs), spin-torque transfer magnetoresistive RAMs (STT-MRAMs), and resistive RAMs (RRAMs). Table I shows a comparison of emerging NVRAM technologies reported by Meena et al. [1], Burr et al. [2], Wong et al. [3], with quantum dot access channel nonvolatile random access memories (QDAC-NVRAM). Here, F^2 is the footprint of PCM. The QDAC-NVRAM cells have ~1.55× larger cell size compared to conventional Si-oxide-nitride-oxide-Si (SONOS) flash. This is due to an extra Drain D2 terminal as shown in Figs. 1 and 2. Unlike competing technologies, QDAC-NVRAM is compatible with CMOS processing.

[*]Corresponding author.

Table I. Comparison of emerging NVRAM technologies with QDAC-NVRAM.

Features	FeRAM	MRAM	STT-RAM	PCM	QDAC-NVRAM**
Cell size (F^2)	20-40 F^2	25 F^2	6-20 F^2	F^2	< F^2
Storage mechanism	Polarization of a ferroelectrics (PZT or SBT)	Magnetization of a ferro-magnetic tunnel junctions (MTJ)	Torque on magnetic moment*	Amorphous - polycrystalline phases of GST	Floating gate
Read time (ns)	20 to 80	3 to 20	2 to 20	20 to 50	< 20
Write/erase time (ns)	50/50	3 to 20	2 to 20	20/30	20/30
Endurance	10^{12}	>10^{15}	>10^{16}	10^{12}	TBD
Write power	Mid	Mid to high	Low	Low	Very Low
Nonvolatility	Yes	Yes	Yes	Yes	Yes
Maturity	Limited production	Test chips	Test chips	Test chips	Research Phase
Applications	Low density	Low density	High density	High density	High density

*Spin-polarized current applies torque on the magnetic moment **Estimates based on simulated models

We describe a methodology which transforms a nonvolatile memory, using quantum dot arrays as the floating gate, into a random access nonvolatile memory having comparable write and erase times as competing technologies. This is achieved by incorporating a quantum dot access channel in proximity of the floating gate [4]. The QDAC is biased only during the erase operation and is not biased during read operation.

2. Quantum Dot Access Channel Nonvolatile Random Access Memory (QDAC-NVRAM)

Figure 1 shows the 3-D schematic of a quantum dot nonvolatile RAM, the inset shows the transmission electron microscopy (TEM) image of the self-assembled Si QDs [5]. Four layers of SiO_x cladded Si quantum dots (QDs) (~4nm Si core and ~1nm SiO_x cladding layer) form the floating gate region. The control gate dielectric consists of a ~7.5nm thick layer of silicon nitride upon which a 100nm thick aluminum layer was deposited. In addition to source (S) and drain (D1), a second drain (D2) is introduced adjacent to the gate (G) to serve as an electrical access point to the floating gate. The quantum dots were deposited using a site-specifically self-assembly technique [6,7]. The bottom two quantum dot layers, in proximity of the tunnel oxide, form the floating gate while the top two quantum dot layers form the quantum dot access channel (QDAC) contacted by D2. The QDAC layer can also store the charges tunneled from the bottom QD floating gate layers during the write operation based on the bias potential.

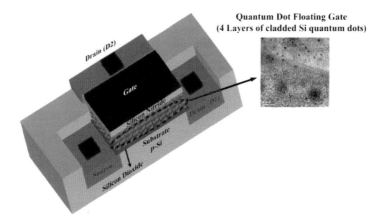

Fig. 1(a). 3-D schematic of a fabricated QDAC-NVRAM with TEM image of cladded Si quantum dot layers in the floating gate [4,7].

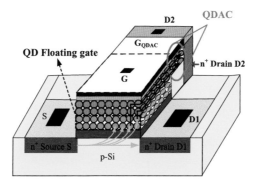

Fig. 1(b). QDAC-NVRAM with thinner control gate dielectric in QDAC (G_{QDAC}) region, accessible with drain D2.

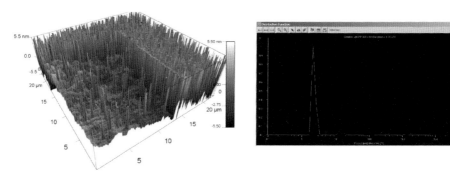

Fig. 1(c). AFM image of self-assembled SiO_x-cladded Si QDs on a patterned p-type Si substrate [7,8].

Fig. 1(d). DLS of SiO_x-Si QDs in the colloidal solution with average particle size (APS) < 6 nm.

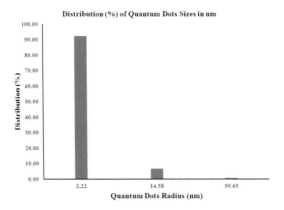

Fig. 1(e). Distribution of SiOx-Si QDs in the supernatant colloidal QD solution of APS < 6nm from DLS is 93.9%.

Figure 1(b) shows the schematic view with quantum dot access channel details. The top two layers of brown colored quantum dots represent the QDAC, which is in proximity of the gray colored QD floating gate layer. The QD floating region is occupied by the tunneled electrons from the inversion channel (shown by the pink arrows/lines) during the write operation. Portion of the QDAC can be used to store the electrons based on the applied G and D1 voltages. The charges can tunnel to the QDAC region when high bias voltages are applied. These tunneled electrons are stored in the QDAC region, which are not affected by drain D2, as D2 is not biased during the write operation. The QDAC behaves like a FET [4,5] with a virtual source region, with the charges stored in the floating gate. The stored charges in floating gate are removed via drain terminal D2 by applying appropriate biasing at the gate and drain D2 (shown by black lines/arrows). The gate insulator thickness over the QDAC section (shown as G_{QDAC} portion) is thinner than the control dielectric thickness in the floating gate region (shown as G). The thicker control gate dielectric under G is needed for charge retention and endurance. However, it increases the write voltage and time period. Thinner control gate dielectric in G_{QDAC} part enables faster erase path at lower voltages/times. The access times can be reduced further by using materials for channel with higher mobility such as cladded Ge QD channel in the QDAC region.

Important physical parameters which can cause the variation across devices are: (i) uniformity of the size of self-assembled QDs, (ii) uniform assembly of QD layers in the floating gate and the QDAC regions, and (iii) QD cladding layer thickness, (iv) thickness of the tunneling which affects the electron tunneling rate from the inversion channel to the QDs in the floating gate. Figure 1(a) shows the TEM image of QDs with uniform sizes. Figure 1(c) shows the atomic force microscopy (AFM) image of uniformly self-assembled two layers on SiO_x-cladded Si QDs on p-type Si substrate [5]. Figure 1(d) shows the size distribution of QDs in supernatant colloidal QD solution with average particle size (APS) of < 6nm from dynamic light scattering (DLS) results. DLS results showed that 93.9% of QD particles are within the desired particle sizes is presented in Fig. 1(e). In addition, the thin SiO_x cladding layer around the QD core enables the formation of quantum dot superlattice (QDSL) which has min-energy bands that can compensate portion of the dot size variation. The tunnel and control gate thickness variation can be controlled well with latest fabrication techniques.

3. QDAC-NVRAM Experimental Characteristics

QDAC-NVRAM write operation can be performed in two ways: (i) connecting the source terminal to ground, applying positive bias voltages at gate (G/G_{QDAC}) and drain D2 terminals, and (ii) connecting the source terminal to ground, applying positive bias to the gate and drain D1 terminals (conventional writing mechanism and requires longer write cycle).

Multi-bit storage in QD-NVRAMs is natural due to the degree of coupling among cladded quantum dots forming an array. Experimental I_{D1}-V_G and I_{D1}-V_{D1} characteristics of a 15 micron × 10 micron QDAC-NVRAM are shown in Figs. 2(a) and 2(b), respectively [7]. Figure 2(a) demonstrates the multi-bit storage capability when write pulses with different voltages and durations are applied across G and D1 terminals. When the applied gate voltage (V_G) is greater than the threshold voltage (V_{TH}), the inversion channel occurs.

The black line represents the I_{D1}-V_G characteristics when no external write pulses were applied and the QDs in the floating gate were not filled with electrons. When write pulse-1 was applied with a gate pulse of 22V and a drain D1 pulse of 10V for 12µs, the electrons started tunneling from the inversion channel to the bottom Si QD layer, QD layer closer to the channel. These tunneled electrons stored in the bottom Si QD layer and the stored carriers increase the device threshold voltage and shift the characteristics shown as the red line. The threshold voltage of the device after applying write pulse-1 is given by ($V_{TH} + \Delta V_{TH\text{-}WP1}$). The maximum voltage separation of 1.5V was observed relative to no write pulses (black line) at 300µA of I_D. The shift is towards the right is highlighted with the red arrow.

When the pulse voltages were increased further to write pulse-2 (24V gate pulse and 10V drain D1 pulse for 12µs), the electrons tunnel to the second Si QD layer from the bottom after all the QDs in the bottom layer were completely filled. As the tunneled electrons in the middle QD layers contributed to further change in the threshold voltage shift ($\Delta V_{TH\text{-}WP2}$) results in ($V_{TH} + \Delta V_{TH\text{-}WP2}$). The capacitance of the floating gate ($C_{CG\text{-}FG}$) is smaller when the electrons are only in the bottom QD layer than the capacitance when the electrons are in the middle QD layers. This is due to the reduction in the effective insulator thickness between the control gate and the QD layer with charges. The increase in $C_{CG\text{-}FG}$ results in reduced $\Delta V_{TH\text{-}WP2}$ ($< \Delta V_{TH\text{-}WP1}$). The effect of additional charges stored in the floating gate results in higher voltages are required to make the transition to middle QD layers from the bottom QD layer. Due to smaller $\Delta V_{TH,}$ the characteristics move to the left, shown as the blue line, when compared to electrons stored only in the bottom QD layer (red line). The blue arrow in Fig. 2(a) represents the shift.

With increasing write pulse voltages and durations, the electrons start filling the vacant middle and upper Si QD layers. The electrons tunnel and fill the vacant places in the remaining QD layers with the application of write pulse-3, pulse-4, pulse-5, and pulse-6. The applied Write pulses resulted in further reduction in threshold voltage shifts (ΔV_{TH}) respectively. The characteristics shift left corresponding to write pulse-3, pulse-4, pulse-5, and pulse-6 and are presented by green, pink, gold and orange lines respectively.

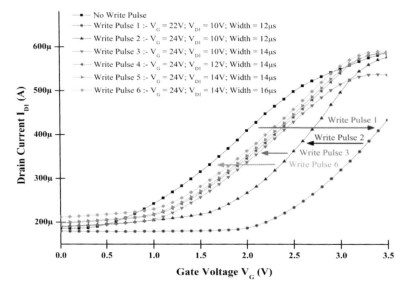

Fig. 2(a). Experimental I_{D1}-V_G characteristics of a QDAC-NVRAM showing multi-bit storage capability.

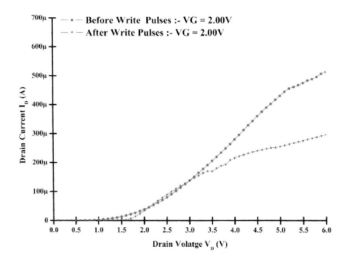

Fig. 2(b). Experimental I_{D1}-V_{D1} characteristics of a fabricated 15μm × 10μm device before and after the application of write pulses.

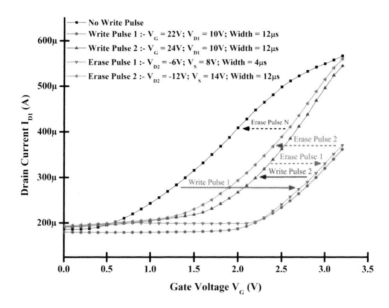

Fig. 2(c). Experimental I_{D1}-V_G Characteristics showing the charge retrieval through the source terminal.

As the write pulse voltage and duration increases, the electrons transfer within the upper QD layers closer to the gate. From characteristics in Fig. 2(a), we observe that with the application of more write pulses, the threshold voltage shift and the voltage separation between the individual bits get smaller and become difficult to realize them as separate bits. The maximum threshold voltage shift is observed when the electrons tunnel from the inversion channel to the bottom QD layer, indicated with the red arrow, and then with increased pulsing conditions lead to reduction in the ΔV_{TH}.

Figure 2(b) presents the experimental I_{D1}-V_{D1} characteristics of a QDAC-NVRAM before and after the write pulses were applied. Due to the stored charges in the floating gate, the threshold voltage of the device is higher when compared to no write pulses were applied, which led to the reduction of I_{D1}.

QDAC-NVRAM erase operation can also be performed in two ways: (i) applying negative bias to the source, and apply positive bias to drain D2 and gate, and (ii) applying positive potential to the source and gate terminals, and applying negative potential to the drain D2.

In Fig. 2(c), black line represents the characteristic with no charges stored in the floating gate. When the write pulses '1' and '2' were applied, the write bits are represented by the red and blue characteristics [7]. When the erase pulse-1 of 8V/4μs at the source and -6V/4μs at the drain D2 are applied, the characteristics moved from the blue line to the green line and the shift is highlighted with the green arrow. This indicates that the electrons were removed due to write pulse-2 i.e., from the second QD layer in the floating gate. When erase pulse-2 with voltage/duration of 14V/12μs was applied at the source and -12V/12μs at the drain D2, the characteristic moved from the green characteristic to the pink. This indicates that a portion of the charges stored in the bottom QD layer were removed. As the magnitude/duration of erase Pulses is increased further, the characteristics moves towards the black characteristics, which represents no electrons in the QD floating gate.

4. QDAC-NVRAM Simulation Integrating QD Nonvolatile Memory (QD-NVM) and QDAC Models

When the external gate voltage pulses are higher than the threshold voltage of the device, the electrons start to tunnel from the inversion channel to the bottom or the top of QD layers in the floating gate region. The threshold voltage shift varies based on the electron tunneling occurs between the inversion channel and either to the top or to the bottom layer of QDs in the floating gate during the write/read and erase operations. The threshold voltage of a QDNVM, $V_{TH-QDNVM}$ is dependent on the amount of charge present in the QD floating gate is given in Eq. (1) [6,8-10].

$$V_{TH-QDNVM} = V_{TH} + \Delta V_{TH} \tag{1}$$

Where V_{TH} is the threshold voltage expressed by Eq. (2)

$$V_{TH} = V_{FB} - \frac{Q_{sc}}{C_{ox}} + 2\psi_B = \left(\phi_{ms} - \frac{Q_{ox}}{C_{ox}}\right) - \frac{Q_{sc}}{C_{ox}} + 2\psi_B \tag{2}$$

Here in Eq. (2), V_{FB} is the flat-band voltage of a MOSFET, Q_{sc} is the surface charge within the depletion region per unit area, C_{ox} is the tunnel oxide capacitance per unit area, ψ_B is the energy difference between the intrinsic Fermi level (E_i) and the Fermi level (E_F), ϕ_{ms} is the work-function difference between the gate metal and the semiconductor and Q_{ox} is the fixed-oxide charge per unit area. The threshold voltage shift (V_{TH}) due to the charges present in the QDs can be expressed as Eq. (3) [9]

$$\Delta V_{TH} = \frac{Q}{C} = \frac{\int_0^{t_w} j(t) A dt}{C_{CG-FG}} \tag{3}$$

Where $j(t)$ is the current per unit area flowing during the charging time (t_w) of the QD floating gate while performing the write or program operation which is given in Eq. (4) below

$$j(t) = q * n_{dot} * N_{QD} * P_{w \rightarrow d} \tag{4}$$

Here, q is the charge of the electron, n_{dot} is the number of electrons per dot (bound as well as the electrons trapped at the QDs interface), N_{QD} is the density of QDs, and P is the tunneling rate of carriers from the channel (quantum well) to the quantum dots. C is the capacitance between the control gate and the QD floating gate which is expressed in Eq. (5)

$$C_{CG-FG} = \frac{\varepsilon_{CD}\varepsilon_0 A}{t_{CG-FG}} \quad (5)$$

The tunneling rate between inversion channel (well W) and QDs (d), $P_{w \to d}$ is computed by Eq. (6) [13]

$$P_{w \to d} = \frac{4\pi}{\eta} \sum_{w,d} |\langle \psi_d | H_t | \psi_w \rangle|^2 (f_w - f_d) \delta(E_d - E_w) \quad (6)$$

Here, ψ_d and ψ_w are the wavefunctions in the quantum dot and the quantum well respectively, f_d and f_w are the Fermi distribution functions of the dot and the well, H_t is the Hamiltonian, E_w and E_d are the energy levels in the inversion channel and the dots. The amount of the charges transferred to the QDs depends on the tunneling probability of the wavefunctions of the inversion channel (ψ_w) and the quantum dots (ψ_d). The charge in a quantum dot in a layer can be calculated by using the electrons tunneling rate from the channel to the dots. The electron distribution in an inversion channel/quantum well is calculated by solving the 1-D Schrödinger's and the Poisson's equations self-consistently as described above. The model will be improved to simulate sub-12 nm devices.

5. QDAC Transport Model

Recently, we have successfully computed the density of states in various energy minibands [5,11]. This permits computing the FET current in QDC-FETs and finding the erase time.

$$I_{D,QDAC} = \left(\frac{W}{L}\right) C_o'' \left[\sum_{i=0}^{n} \mu_{ni} \left\{ V_G - (V_{THi} + \Delta V_{THi}) - \frac{1}{2} V_{DS} \right\} V_{DS} \right] \quad (7)$$

Eq. (7) is an empirical relation expressing the influence of gate voltage accounting for the carriers in energy mini-bands (i). Here, the threshold voltage V_{THi} and shift ΔV_{THi} depend on the concentration of electrons (dependent of V_G) which in turn determines the occupancy of number of energy mini-bands.

6. QDAC-NVRAM Model

The QD-NVM model developed by Hasaneen, Heller and Jain [9] will be modified to incorporate QDAC-enabled charge transport during erase operations. The threshold voltage shift (ΔV_{TH}) in QD-NVM is due to stored charges in the floating gate, which is dynamically modified by electrons transferring from the inversion channel through tunnel oxide/insulator due to Fowler-Nordheim tunneling during write operation. Cladded QDs improve ΔV_{TH} over NVMs using non-uniform dots without cladding layer.

108 M. Lingalugari et al.

Using Eqs. (3) and (4), ΔV_{TH} is expressed as:

$$\Delta V_{TH} = Q_{FG}/C_{CG-FG} = (1/C_{CG-FG})\int_0^{t_w}(q*n_{dot}\,N_{QD}*P)A\,dt. \quad (8)$$

Here, N_{QD} is the density of QDs, n_{dot} is the number of electrons per dot, C_{CG-FG} is the capacitance between the control gate and quantum dots, and P is the tunneling rate of carriers from the channel to QDs.

Figure 3(a) shows simulation of electrons in the inversion channel at a gate voltage $V_G = 1.0V$ for a two cladded SiO_x-si quantum dot layers serving as the floating gate, sandwiched between tunnel oxide and control gate dielectric Si_3N_4. As V_G is increased to 1.2V part of electrons from the channel tunnel to the lower Si QD layer [Fig. 3(b)]; subsequent increase in V_G (not shown) transfers some electrons to the upper Si QD layer. The simulation parameters are listed in the Table II. Here, electron affinity is χ, electron and hole effective masses are m_e-m_h, doping levels are N_a and N_d, energy band gap E_g, and dielectric constant ε_r.

Figure 3(c) shows the simulated I_D-V_D characteristics of a 140nm × 140nm (0.14 μm × 0.14 μm) NVRAM cell, based on a BSIM equivalent device model. The estimated erase time is less than ~90ns with a threshold voltage shift of 0.3V, based on extrapolation from the experimental results. The erase time of the QDAC-NVRAM is calculated using the driving current I_D (QDAC) as expressed in Eq. (7) using the parameters of the QDAC channel connected to drain D2. Here we operate at much lower voltages and time constants during erase which is done through QDAC by applying a positive voltage pulse to D2. In this scheme appropriate gate voltage (V_G) is applied to turn ON the QDAC (a QDC-FET with floating gate serving as source). These erase times can be reduced below 50ns by employing QDAC with high mobility GeO_x-Ge quantum dots. The erase time of a scaled floating gate device with dimensions 14nm × 14nm is approximately 5ns to ~500 ps range. Figure 3(d) shows the effect of quantum dot cladding thickness varying from 5Å to 20Å on tunneling time of electrons from inversion channel to the quantum dot

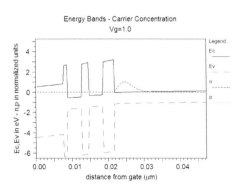

Fig. 3(a). Charge (dotted red peak) in the inversion channel of the QD-NVM.

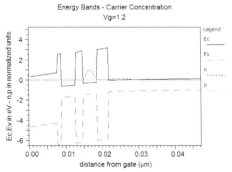

Fig. 3(b). Charge transfer to bottom Si QD layer from the inversion channel, when 1.2V at applied the gate.

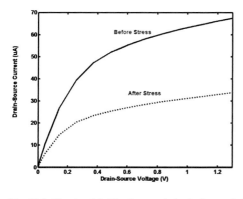
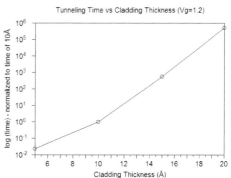

Fig. 3(c). Simulated I_D-V_D characteristics before and after Write pulses for 140nm x 140nm memory.

Fig. 3(d). Tunneling time variation as a function of the QD cladding layer thickness varying from 5Å to 20Å.

Table II. Parameters used for the quantum simulation of Figs. 3.

Layer	Thick (μm)	χ (eV)	E_g (eV)	m_e	m_h	ε_r	N_d (cm^{-3})	N_a (cm^{-3})
Si$_3$N$_4$-control gate insulator	0.0075	2.7	5.0	0.16	0.16	7.5	0.0e00	0.0e00
SiO$_x$ QD cladding	0.0010*	0.9	9.0	0.5	0.5	3.9	0.0e00	0.0e00
Si QD core	0.0040	4.15	1.12	0.19	0.49	11.9	0.0e00	0.0e00
SiO$_x$ cladding	0.0020*	0.9	9.0	0.5	0.5	3.9	0.0e00	0.0e00
Si QD core	0.0040	4.15	1.12	0.19	0.49	11.9	0.0e00	0.0e00
SiO$_x$ QD cladding	0.0010*	0.9	9.0	0.5	0.5	3.9	0.0e00	0.0e00
SiO$_2$-tunnel insulator	0.0020	0.9	9.0	0.5	0.5	3.9	0.0e00	0.0e00
Si substrate	0.5000	4.15	1.12	0.19	0.49	11.9	0.0e00	1.0e16

floating gate. From the simulations, we observe that we can optimize the threshold voltage shift, write time, and operating voltages by engineering the tunnel oxide and QD cladding parameters.

7. Memory Array Architecture for QDAC-NVRAMs with Dedicated Write/Erase Line

Figure 4(a) shows the top view of one QDAC-NVRAM cell layout terminals electrical connections (bit-line, word line, and dedicated erase/write line) in a memory array architecture proposed in Fig. 4(b). QDAC-NVRAM size (12λ × 14λ) is larger than the conventional flash memory cell (12λ × 9λ) due to the additional drain D2.

Conceptual QDAC-NVRAMs based 4 × 4 memory array architecture is shown in Fig. 4(b). The gates (G) of the memory cells on each row of the array are connected to a word line (WL). The sources of the cells in array are connected to supply line. The drains (D1) of the cells on each row are connected to a bit line (BL). Similarly, the drains (D2) of the cells on each row are connected to a dedicated write/erase line.

110 M. Lingalugari et al.

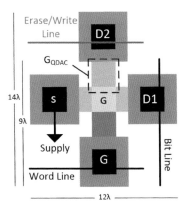

Fig. 4(a). Layout of one QDAC-NVRAM cell with D1, G, D2, and S connected to bit line, word line, write/erase line, and ground respectively. The W/L ratio of gate is 10μm/14μm, source and drain D1 extensions are 6μm × 20μm, and drain D2 extension is 14/20.5μm. The source and drain D1 regions (blue) are 50μm × 60μm, and drain D2 (blue) is 30μm × 45μm. The source and drain D1 contacts (black) are 30μm × 30μm, and drain D2 (black) contact 20μm × 35μm.

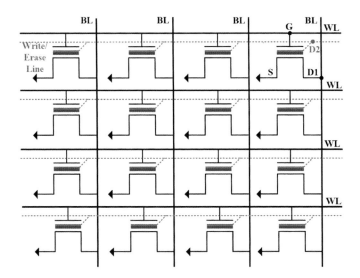

Fig. 4(b). QDAC-NVRAM based 4×4 array architecture with dedicated write/erase line.

During the write operation, individual cells are accessed by applying voltage pulses with given amplitudes but different widths to WL and BL. During read operation, the individual cells are selected similar to write operation. However, the voltage pulses applied to the WL are lower in amplitude. If a selected memory cell stores a "0", its BL will be pulled down and read out by the sensing amplifier at the end of the bit line. Data "1" or "2" cannot be read by this narrow word line pulse due to their low drain currents;

the bit line will only be pulled down slight but cannot be sensed by the amplifier. Then, a word line pulse with a wider width is applied to detect if the memory cell stores a "2" (data "2" has a larger drain current than data "1" but a smaller drain current than data "0"). If the sensing amplifier detects the sufficient pull-down on the bit line, the memory cell stores a "2"; otherwise it stores a "1". As a result, it needs two read cycles to read the possible three values in a memory cell.

During erase operation, the individual cells are selected similar to write operation. The novelty of this approach is the dedicated write/erase line which directly accesses the floating gate charge via QDAC and drain D2 terminals. Further experiments are needed to determine the robustness and speed of this memory array. Biasing of write/erase line with respect to source S and gate G determines the charge removal mechanism. Even though we have an additional erase line, this helps in high-speed erasing and thus transforming the nonvolatile memories to high-speed and low-power random access memories [1-3].

During erase: Voltages: V_{D2} and V_G (QDAC-NVRAM) $<<$ $V_{SG,FN}$ (conventional NVMs), Time constants: $R_{eq}C$ (QDAC-NVRAM) $<<$ $R_{eq}C$ (conventional NVMs).

This comparison is for similar electric field across the tunnel insulator. The erase time of QDAC-NVRAM is estimated to be comparable to phase change (PCMs), MRAM, and RRAMs [1-3].

8. Conclusion

We have presented write/erase characteristics of fabricated long-channel (W/L = 15μm/10μm) nonvolatile random access memories (NVRAMs) with 4μs erase times. These devices are CMOS-compatible and employ novel quantum dot access channel (QDAC) which enables fast storage and retrieval of charge from the floating gate layer by directly accessing via an additional drain D2. In addition, we have demonstrated multi-bit storage and erase of a QDAC-NVRAM. QDAC-NVRAMs have potential scalable to sub 22nm. We also proposed memory array architecture using the novel QDAC-NVRAMs. QDAC-NVRAMs demonstrate comparable erase times to competing technologies. Unlike, other competing technologies QDAC-NVRAMs can be fabricated using conventional CMOS processes.

Acknowledgment

The authors gratefully acknowledge Dr. T.P. Ma and his laboratory personnel at Yale University for their assistance in memory testing. Authors also would like to acknowledge Dr. Ronald LaComb for his assistance in the device fabrication using CNS facilities at Harvard University.

References

[1] J.S. Meena, S M. Sze, U. Chand, and T-Y. Tseng, Overview of emerging nonvolatile memory technologies, Nanoscale Res. Lett., 9, p. 526, 2014.

[2] G.W. Burr, B.N. Kurdi, J.C. Scott, C.H. Lam, K. Gopalakrishnan and R.S. Shenoy, "Overview of Candidate Device Technologies for Storage-Class Memory", IBM J. res. & Dev., Vol. 52, No. 4/5, July/September, 2006.

[3] H.S.P. Wong, S. Raoux, S.B. Kim, J. Liang, J.P. Reifenberg, B. Rajendran, M. Asheghi and K.E. Goodson, "Phase Change Memory", Proceedings of The IEEE, 2010.

[4] F. Jain, Nonvolatile memory and three-state FETs using cladded quantum dots, US Patent 9153594, October 6, 2015.

[5] F. Jain, S. Karmakar, P.-Y. Chan, E. Suarez, M. Gogna, J. Chandy, and E. Heller, "Quantum Dot Channel (QDC) Field-Effect Transistors (FETs) Using II-VI Barrier Layers", J. Electronic Materials, 41, 10, pp. 2775-2784, August 2012.

[6] R. Velampati, E-S. Hasaneen, E. K. Heller and F. C. Jain, "Floating Gate Novel Nonvolatile Memory Using Individually Cladded Monodispersed Quantum Dots", IEEE Trans. VLSI, 25, 5, 2017. DOI: 10.1109/TVLSI.2016.2645795.

[7] M. Lingalugari, Quantum Dot Based Multi-Bit FETs and Memories, PhD Thesis, University of Connecticut (2016).

[8] M. Gogna, E. Suarez, P.Y. Chan, F. Al-Amoody S. Kamakar and F. Jain, "Nonvolatile Silicon Memory using GeO_x-Cladded Ge Quantum Dots Self-Assembled on SiO_2 and Lattice-Matched II-VI Tunnel Insulator", in Journal of Electronic Materials, Vol. 40, pp. 1769-1774, Aug. 2011.

[9] E-S. Hasaneen, E. Heller, R. Bansal, and F. Jain, "Modeling of Nonvolatile Floating Gate Quantum Dot Memory", Solid-State Electronics, 48, pp. 2055-2059, September 2004.

[10] F.C. Jain, E. Suarez, M. Gogna, F. Alamoody, D. Butkiewicus, R. Hohner, T. Liaskas, S. Karmakar, P-Y. Chan, B. Miller, J. Chandy, and E. Heller, "Novel Quantum Dot Gate FETs and Nonvolatile Memories Using Lattice-Matched II-VI Gate Insulators", J. Electronic Materials, 38, pp. 1574-1578, August 2009.

[11] F. Jain, M. Lingalugari, J. Kondo, P. Mirdha, E. Suarez, J. Chandy, and E. Heller, Quantum dot channel (QDC) FETs with Wraparound II-VI Gate Insulators: Numerical Simulations, J. Electronic Materials, 45, 11, pp. 5663-70, November 2016.

[12] K. Itoh, An introduction to memory chip design, Springer Series in Advanced Microelectronics, Volume 5, 2001, pp. 1-48 (Springer-Verlag, NewYork).

[13] S. Yuasa, A. Fukushima, K. Yakushiji, T. Nrozaki, M. Konoto, H. Maehara, H. Kaubota, T. Taniguchi, H. Arai, H. Imamura, K. Ando, Y. Shiota, F. Bonell, Y. Suzuki, N. Shimomura, E. Kitagawa, J. Ito, S. Fujita, K. Abe, K. Nomura, H. Noguchi, and H. Yoda, Future prospects of MRAM technologies, Proc. IEDM, pp. 3.1.1-3.1.4, 2013.

[14] Y.Y. Chen, M. Komura, R. Degraeve, B. Govoreanu, L. Goux, A. Fantini, N. Raghavan, S. Clima, L. Zhang, A. Belmonte, A. Redolfi, G.S. Kar, G. Groeseneken, D. Wouters, and M. Jurczak, Improvement of data retention in HfO_2/Hf 1T1R RRAM cell under low operating current, Proc. IEDM, pp. 10.1.1-10.1.4, 2013.

[15] C-W Hsu, C-C. Wan, I-T. Wang, M-C. Chen, C-L. Lo, Y-J. Lee, W-Y. Jang. C-H. Lin, T-H. Hou, 3D vertical TaO_x/TiO_2 RRAM with over 10^3 self-rectifying ration and sub-μA operating current, Proc. IEDM, pp. 10.4.1-10.4.4, 2013.

[16] S. Rusu, S. Tam, H. Muljono, D. Ayers, J. Chang, R. Varada, M. Ratta, and S. Vora, A 45nm 8-core enterprise Xeon® processor, A-SSCC, pp. 9-12, 2009.

[17] C. Kim, and L. Chang, "Guest Editors' Introduction: Nanoscale Memories Pose Unique Challenges", IEEE Design & Test, 2011.

Quantum Dot Channel (QDC) Field Effect Transistors (FETs) Configured as Floating Gate Nonvolatile Memories (NVMs)

Jun Kondo[*,‡], Murali Lingalugari[*], Pial Mirdha[*], Pik-Yiu Chan[*], Evan Heller[†,¶] and Faquir Jain[*]

[*]Electrical and Computer Engineering,
University of Connecticut,
371 Fairfield Way, Unit 4157, Storrs, CT 06269, USA
[†]Synopsis Inc., Ossining, NY 10562, USA
[‡]jun.kondo@uconn.edu
[¶]evankheller@gmail.com

This paper presents quantum dot channel (QDC) Field Effect Transistors (FETs) which are configured as nonvolatile memories (NVMs) by incorporating cladded GeO_x-Ge quantum dots in the floating gates as well as the transport channels. The current flow and the threshold characteristics were significantly improved when the gate dielectric was changed from silicon dioxide (SiO_2) to hafnium aluminum oxide ($HfAlO_2$), and the control dielectric was changed from silicon nitride (Si_3N_4) to hafnium aluminum oxide ($HfAlO_2$). The device operations are explained by carrier transport in narrow energy mini-bands which are manifested in a quantum dot transport channel.

Keywords: quantum dot; transport channel; gate dielectric; control dielectric; mini-band.

1. Introduction

As CMOS device dimensions continue to shrink in new integrated circuit technology, the gate dielectric thickness also continues to shrink, and gate leakage becomes a major issue.[1,2] The gate dielectric thickness of SiO_2 had been reduced below 30Å by the year 2000, and there was a concern about further reduction of the gate oxide thickness.[2] Therefore, an alternative gate dielectric material must be found in order to replace the traditional silicon dioxide (SiO_2) gate dielectric to prevent gate leakage.[1] For a given equivalent oxide thickness (EOT), using a higher dielectric constant and a larger physical thickness yields significant suppression of direct tunneling gate current.[2] Silicon dioxide (SiO_2) and silicon nitride (Si_3N_4) have the dielectric constants of 3.9 and 7,[3] the band gaps of 9eV and 5.3eV.[3] Because Si_3N_4 has the much higher dielectric constant than SiO_2, numerous attempts were made to make Si_3N_4 the gate dielectric to replace SiO_2, but they have not met with success.[1] On the other hand, the high-k dielectric materials such as tantalum oxide (Ta_2O_3), aluminum oxide (Al_2O_3), titanium oxide (TiO_2), hafnium oxide (HfO_2), and zirconium oxide (ZrO_2) have been also considered as candidates for the gate

[‡]Corresponding author.

dielectric.[4] Among them, Al_2O_3 and HfO_2 have the dielectric constants of 9 and 25,[3] and the band gaps of 8.8 and 5.8.[3] Because Al_2O_3 and HfO_2 have these compensative characteristics, they are combined to produce hafnium aluminum oxide ($HfAlO_2$) which has a relatively large dielectric constant.[5] This paper presents the application of $HfAlO_2$ nanolaminate high-k dielectric combinational layers used for the gate dielectric and the control dielectric of the quantum dot channel floating gate nonvolatile memory.

2. Quantum Dot Channel (QDC) Floating Gate Nonvolatile Memories (NVMs)

The structure of a long channel QDC floating gate nonvolatile memory is shown in Fig. 1. The fabrication of the nonvolatile memory involved the following nine major steps. First, the 1250Å silicon dioxide was deposited on the p-type silicon substrate using the wet oxidation. Second, the source and drain were fabricated on the p-type silicon substrate using the phosphorus diffusion. Third, the gate was opened, and the p-type silicon substrate was exposed at the gate area. Fourth, the recessed region between the n^+ source and the n^+ drain was fabricated using the dry oxidation and the buffered oxide etch (BOE). Fifth, two layers of GeO_x-cladded Ge quantum dots were self-assembled for the transport channel at the recessed region between the n^+ source and the n^+ drain.[6] Electrons flow from the source to the drain through the transport channel. Sixth, twelve $HfAlO_2$ nanolaminate high-k dielectric combinational layers were deposited for the 26Å gate dielectric. Seventh, two layers of GeO_x-cladded Ge quantum dots were self-assembled for the floating gate.[6] If external voltage is applied to the floating gate, the charge stored in the floating gate causes the threshold voltage shift. Eighth, thirty-four $HfAlO_2$ nanolaminate high-k dielectric combinational layers were deposited for the 53Å control dielectric on the floating gate. Finally, a thin aluminum layer was deposited above the source, the drain, and the control dielectric, and divided into three independent electrical terminals for the source, the drain, and the gate. These terminals were used for the I_D-V_G and I_D-V_D characteristic measurements. The right inset in Fig. 1 shows that this nonvolatile memory has the following materials for the gate contact, the control dielectric, the floating gate, the gate dielectric, and the transport channel; aluminum, $HfAlO_2$, GeO_x-cladded Ge quantum dots, $HfAlO_2$, and GeO_x-cladded Ge quantum dots. The $HfAlO_2$ nanolaminate high-k dielectric combinational layers were deposited using the Savannah Atomic Layer Deposition (ALD) system. One cycle of the program execution was involved in separate depositions of Al_2O_3 and HfO_2, and the program was executed twelve cycles to deposit the 26Å gate dielectric, and thirty-four cycles to deposit the 53Å control dielectric. The trimethylaluminum (TMA) was used for the precursor for aluminum oxide (Al_2O_3), and the tetrakis(dimethylamino)hafnium (TDMAH) was used for the precursor for hafnium oxide (HfO_2). The inner heater temperature was setup to 200°C which was below the maximum tolerable temperature 350°C of germanium quantum dots. The nonvolatile memory has a W/L ratio of 26/25.

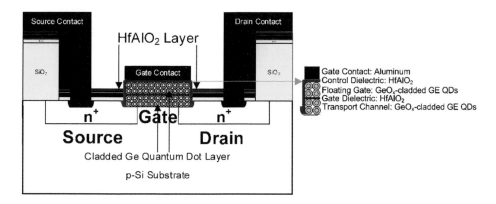

Fig. 1. Structure of the GeO$_x$-cladded Ge quantum dot gate (QDG) quantum dot channel (QDC) floating gate nonvolatile memory (NVM) with the HfAlO$_2$ combinational layers at the tunnel oxide and the control dielectric.

3. Quantum Theory

The Kronig-Penney model was used to determine the energy mini-band locations and their widths in the quantum dot superlattice (QDSL) formed in the array of thin-barrier (~1nm) cladded Ge quantum dots, and the simulated results are shown in Fig. 2.[7,8] The electron transport in the inversion channel is influenced by the energy mini-bands manifested due to the quantum dot channel.

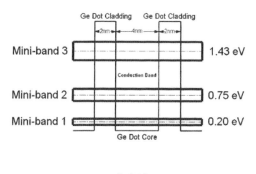

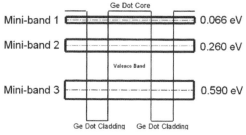

Fig. 2. Energy mini-band Locations in the Ge dot QDSL.[7,8]

The drain current I_D is empirically expressed by Eq. (1).[8] It depends on the number of mini-bands (i) and V_{DSj}, (signified by integer j). The number of mini-bands is determined by the value of V_{DSj} for a given V_G which determines the overall electron charge in the quantum dot channel. Therefore, the onset of various mini-bands can be simulated as if the device as various threshold voltages V_{THi}. The threshold shift ΔV_{TH} depends on the transfer of charge to the QDs in the gate region in QDG-QDC FETs. Here, the quantum dot superlattice forms in quantum dot layers in the gate region.[9]

$$I_D = \left(\frac{W}{L}\right) C_o'' \mu_n \left[\sum_{i=0}^{m}\sum_{j=0}^{n}\left\{V_G - (V_{THi} + \Delta V_{THi}) - \frac{1}{2}V_{DSj}\right\}V_{DSj}\right] \quad (1)$$

The saturation current I_D (sat) is expressed by Eq. (2).[8]

$$I_D(sat) = \frac{1}{2}\left(\frac{W}{L}\right) C_o'' \mu_n \left[\sum_{i=0}^{m}\{V_G - (V_{THi} + \Delta V_{THi})\}\right]^2 \quad (2)$$

4. QDC-Enabled QD Floating Gate Nonvolatile Memory (QDC-NVM)

The quantum dot channel (QDC) field effect transistors (FETs) are configured as floating gate nonvolatile memories (NVMs) by incorporating cladded GeO_x-Ge quantum dots in the floating gates, and depositing thirty-four $HfAlO_2$ nanolaminate high-k dielectric combinational layers on the floating gates for the control dielectrics. During the write operation of the memory, a gate voltage pulse and a drain voltage pulse are applied simultaneously to transfer charge from the inversion channel to the Ge quantum dot layers at the floating gate. The charge in the floating gate shifts the threshold voltage once the 'Write' operation is carried out. The threshold voltage (V_{TH}) of a quantum dot floating gate nonvolatile memory (QD-NVM), $V_{TH-QDNVM}$ is dependent on the amount of charge present in the QD floating gate, and expressed as Eq. (3).[10-16]

$$V_{TH-QDNVM} = V_{TH} + \Delta V_{TH} \quad (3)$$

The threshold voltage shift (ΔV_{TH}), which is due to the charges present in the QDs, is expressed as Eq. (4).[13] C is the capacitance between the control gate and the QD floating gate.

$$\Delta V_{TH} = \frac{Q}{C} = \frac{\int_0^{t_w} j(t) A dt}{C} \quad (4)$$

Here, $j(t)$ is the current per unit area flowing during the charging time (t_w) of the QD floating gate while performing the write or program operation, and expressed as Eq. (5).

$$j(t) = q * n_{dot} * N_{QD} * P \quad (5)$$

Here, q is the charge of the electron, n_{dot} is the number of electrons per dot (bound as well as the electrons trapped at the QDs interface), N_{QD} is the density of QDs, and P is the tunneling rate of carriers from the channel (quantum well) to the quantum dots. The tunneling rate between inversion channel (shown as well w) and QDs (shown as dot d), $P_{w \to d}$ is expressed as Eq. (6).[10]

$$P_{w \to d} = \frac{4\pi}{\hbar} \sum_{w,d} |\langle \psi_d | H_t | \psi_w \rangle|^2 (f_w - f_d) \delta(E_d - E_w) \qquad (6)$$

Here, ψ_d and ψ_w are the wavefunctions in the quantum dot (in floating gate) and the quantum well (in transport channel), H_t is the Hamiltonian, f_d and f_w are the Fermi distribution functions of the dot and the well, E_w and E_d are the energy levels in the inversion channel and the dots. Therefore, the amount of the charges transferred to the QDs depends on the tunneling probability of the wavefunctions of the inversion channel (ψ_w) and the quantum dots (ψ_d). The charge in a quantum dot in a layer can be calculated by using the electron tunneling rate from the channel to the dots.[15]

5. Quantum Simulation

The electron distribution in a transport channel/quantum well is calculated by solving the 1-D Schrödinger's and the Poisson's equations self-consistently. In this simulation, the gate dielectric thickness of 50Å, and the control dielectric thickness of 20Å are used. Table 1 shows the parameters used for the simulation of device characteristics for the QDC NVM, and Table 2 shows the parameters used for the simulation of the Ge QDC NVM with HfAlO$_2$ combinational layers. Figure 3(a) shows the simulation of charge transfer to Ge QDs, 'Writing' bit '1', and a part of electrons is migrated from the inversion channel to the lower Ge QD layer of the floating gate. Figure 3(b) shows the charge density in the transport channel, and the plateau part of the charge density line indicates that there is onset of tunneling from the transport channel to the floating gate.

6. Experimental Results

The measured I_D-V_G and I_D-V_D characteristics of the fabricated floating gate nonvolatile memory are shown in Fig. 4(a) and Fig. 4(b). The drain pulse of 4V for 400μs and the gate pulse of 15V for 100μs were applied simultaneously. In Fig. 4(a) and Fig. 4(b), the labels B4, B5 and B6 correspond the drain and gate voltages of 4, 5, and 6V before the pulse, and the labels A4, A5 and A6 correspond the drain and gate voltages of 4, 5, and 6V after the pulse. The maximum threshold voltage shift ΔV_{TH} in I_D-V_G characteristics was approximately 0.9V at the drain current of 246μA. The maximum threshold voltage shift ΔV_{TH} in I_D-V_D characteristics was approximately 0.8V at the drain current of 390μA. The maximum current flow in I_D-V_G characteristics was 522μA before the pulse, and 246μA after the pulse. The maximum current flow in I_D-V_D characteristics was 704μA before the pulse, and 390μA after the pulse. On the other hand, I_D-V_G characteristics of the GeO$_x$-cladded Ge and SiO$_x$-cladded Si quantum dot gate (QDG) nonvolatile memory are shown before and after the pulses in Fig. 5.[17] For the first test, the drain pulse of 3V and the gate pulse of 11.7V were applied simultaneously for 1 second, and the threshold voltage shift was 0.5V at the drain current of 4μA.[17] For the second test, the drain pulse of 6V and the gate pulse of 30V were applied simultaneously for 1 second, and the threshold voltage shift was -0.2V at the drain current of 4μA.[17] In this device, approximately 60Å of SiO_2 was used for the gate dielectric, and approximately 75Å of Si_3N_4 was used for the control dielectric.[17]

Table 1. Parameters used for the simulation of device characteristics for QDC NVM.

Parameter	Unit	Value	Note
W	μm	60	
L	μm	60	
v_1	cm²/V-s	80	State 1
v_2	cm²/V-s	96	State 2 ($1.2 \times v_1$)
ε_{eff}	—	3.9	Silicon
ε_0	F/cm	8.854×10^{-14}	
t	cm	1.2×10^{-6}	
C_{ox}	F/cm²	8.854×10^{-7}	

Table 2. Parameters used for the simulation of Ge QDC NVM with the $HfAlO_2$ combinational layers.

Layer	Thickness (μm)	χ (eV)	E_g (eV)	m_e	m_h	e_r
$HfAlO_2$	0.002	0.12	7.15	0.10	0.20	17.0
GeO_x	0.0010	2.25	5.70	0.16	0.16	12.0
Ge QD	0.0040	4.55	0.67	0.08	0.28	16.0
GeO_x	0.0020	2.25	5.70	0.16	0.16	12.0
Ge QD	0.0040	4.55	0.67	0.08	0.28	16.0
GeO_x	0.0010	2.25	5.70	0.16	0.16	12.0
$HfAlO_2$	0.005	3.12	7.15	0.10	0.20	17.0
GeO_x	0.0010	2.25	5.70	0.16	0.16	12.0
Ge QD	0.0040	4.55	0.67	0.08	0.28	16.0
GeO_x	0.0020	2.25	5.70	0.16	0.16	12.0
Ge QD	0.0040	4.55	0.67	0.08	0.28	16.0
GeO_x	0.0010	2.25	5.70	0.16	0.16	12.0
p-$Si_{(1.0e16)}$	0.5000	4.15	1.12	0.19	0.49	11.9

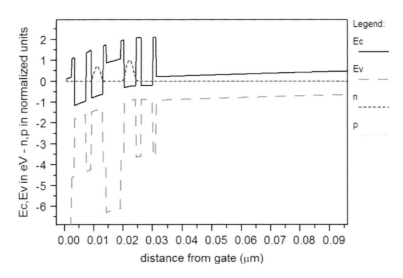

Fig. 3(a). Charge transfer to Ge QDs, 'Writing' bit '1'.

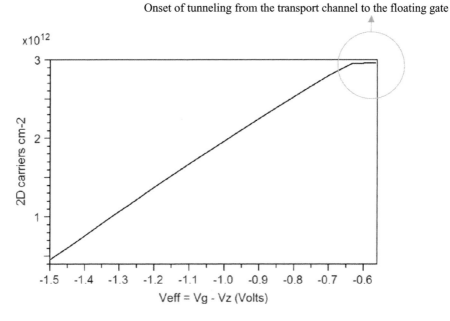

Fig. 3(b). Charge density in the inversion channel.

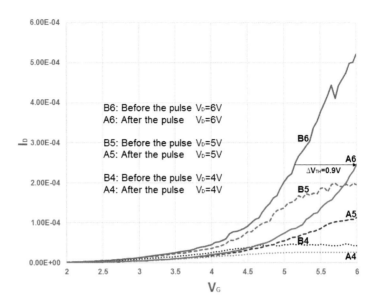

Fig. 4(a). I_D-V_G characteristics of GeO$_x$-cladded Ge quantum dot gate (QDG) quantum dot channel (QDC) floating gate nonvolatile memory (NVM) with the HfAlO$_2$ combinational layers at the gate dielectric and the control dielectric.

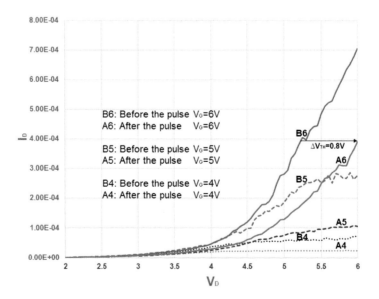

Fig. 4(b). I_D-V_D characteristics of GeO$_x$-cladded Ge quantum dot gate (QDG) quantum dot channel (QDC) floating gate nonvolatile memory (NVM) with the HfAlO$_2$ combinational layers at the gate dielectric and the control dielectric.

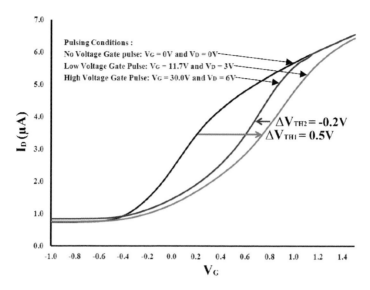

Fig. 5. I_D-V_G Characteristics of the GeO$_x$-cladded Ge and SiO$_x$-cladded Si quantum dot gate (QDG) nonvolatile memory (NVM) with SiO$_2$ at the tunnel oxide and Si$_3$N$_4$ at the control dielectric.[17]

7. Conclusion

When an external voltage is applied to the floating gate for 'Writing' bit '1', the simulation of charge transfer to Ge QDs shows that a fraction of electrons are transferred via tunneling from the inversion channel to the lower Ge QD layer as shown in Fig. 3(a). This is corroborated by simulation of the charge density in the transport channel which shows that there is an onset of tunneling from the transport channel to the floating gate [shown in Fig. 3(b)]. The experimental I_D-V_G characteristics for the fabricated GeO$_x$-cladded Ge QDG-QDC floating gate nonvolatile memory (NVM), having HfAlO$_2$ layers as gate and control dielectric, are presented before and after the 'Write' cycle in Fig. 4(a). In order to find advantages of HfAlO$_2$ nanolaminate high-k dielectric combinational layers over Si$_3$N$_4$, the experimental I_D-V_G characteristics for the fabricated GeO$_x$-cladded Ge and SiO$_x$-cladded Si quantum dot gate (QDG) nonvolatile memory (NVM) with SiO$_2$ at the gate dielectric and Si$_3$N$_4$ at the control dielectric are also presented before and after the 'Write' cycle in Fig. 5. Comparing Fig. 4(a) with Fig. 5, it is concluded that the QDC-FET nonvolatile memory with HfAlO$_2$ nanolaminate high-k dielectric combinational layers at the gate dielectric and the control dielectric provides not only the significantly higher I_D current flow, but also significantly higher threshold voltage shifts which improve the threshold voltage variation, and show the potential for fabricating multi-bit nonvolatile memories.

A comparison of Fig. 4(a), which present ID-VG characteristics of our QDC-NVM with QDG-NVM of Fig. 5[17] shows gradual increase in drain current at a function of gate voltage. QDC-NVM when compared with SONOS (silicon-oxide-nitride-oxide-silicon) shows similar behavior. Additional work is needed to quantitatively compare SONOS Si_3N_4 floating gate performance with QD floating gate.

References

1. P.T. Ma, "Making Silicon Nitride Film a Viable Gate Dielectric," *IEEE Transaction on Electron Devices*, **45**, 3, pp. 680-690 (1998).
2. Y.C. Yeo, Q. Lu, W.C. Lee, T.-J. King, C. Hu, X. Wang, X. Guo, and T.P. Ma, "Direct Tunneling Gate Leakage Current in Transistors with Ultrathin Silicon Nitride Gate Dielectric," *IEEE Electron Device Lett.*, **21**, 11, pp. 540-542 (2000).
3. J. Robertson, "High Dielectric Constant Oxides." Eur. Phys. J. Appl. Phys. **28**, 265-291 (2004).
4. M.-H. Cho, Y. S. Roh, C. N. Whang, and K. Jeong, "Dielectric Characteristics of AL_2O_3-HfO_2 Nanolaminates on Si (100)." Applied Physics Letters, **81**, No. 6, pp. 1071-1073, (2002).
5. P.K. Park, E.-S. Cha, and S.-W. Kang, "Interface Effect on Dielectric Constant of HfO_2/Al_2O_3 Nanolaminate Films deposited by Plasma-enhanced Atomic Layer Deposition." Applied Physics Letters, **90**, No. 232906, pp. 1-3 (2007).
6. F.C. Jain and F. Papadimitrakopoulos, US Patent 7,368,370 B2 (2008).
7. F. Jain, S. Karmakar, P.-Y. Chan, E. Suarez, M. Gogna, J. Chandy, and E. Heller, "Quantum Dot Channel (QDC) Field-Effect Transistors (FETs) Using II-VI Barrier Layers." *J. Electronic Materials*, 41, 10, pp. 2781-2782 (2012).
8. F. Jain, P.-Y. Chan, E. Suarez, M. Lingalugari, J. Kondo, P. Gogna, B. Miller, J. Chandy, and E. Heller, "Four-State Sub-12-nm FETs Employing Lattice-Matched II-VI Barrier Layers." *J. Electronic Materials*, **42**, 11, pp. 3199-3200 (2013).
9. J. Kondo, M. Lingalugari, P.-Y. Chan, E. Heller, and F. Jain, "Modeling and Fabrication of Quantum Dot Channel Field Effect Transistors Incorporating Quantum Dot Gate", Proceedings of TechConnect World 2013, Washington, D.C. (2013).
10. F. C. Jain, E. Suarez, M. Gogna, F. Alamoody, D. Butkiewicus, R. Hohner, T. Liaskas, S. Karmakar, P-Y. Chan, B. Miller, J. Chandy, and E. Heller, "Novel Quantum Dot Gate FETs and Nonvolatile Memories Using Lattice-Matched II-VI Gate Insulators", *J. Electronic Materials*, **38**, 8, pp. 1574-1578 (2009).
11. M. Gogna, E. Suarez, P.-Y. Chan, F. Al-Amoody S. Kamakar, and F. Jain, "Nonvolatile Silicon Memory Using GeO_x-Cladded Ge Quantum Dots Self-Assembled on SiO_2 and Lattice-Matched II-VI Tunnel Insulator", *J. Electronic Materials*, **40**, 8, pp. 1769-1774 (2011).
12. R. Velampati, and F. C. Jain, "A Novel Nonvolatile Memory Using SiO_x-Cladded Si Quantum Dots", NSTI Nanotech, Santa Clara, CA, May 20-24 (2007).
13. E.-S. Hasaneen, E. Heller, R. Bansal, W. Huang, and F. Jain, "Modeling of Nonvolatile Floating Gate Quantum Dot Memory", *Solid-State Electronics*, **48**, pp. 2055-2059 (2004).
14. P.-Y. Chan, M. Gogna, E. Suarez, S. Karmakar, F. Al-Amoody, B.I. Miller, and F.C. Jain, "Nonvolatile Memory Effect in Indium Gallium Arsenide-Based Metal-Oxide-Semiconductor Devices Using II-VI Tunnel Insulators", *J. Electronic Materials*, **40**, 8, pp. 1685-1688 (2011).
15. P.-Y. Chan, E. Suarez, M. Gogna, B.I. Miller, E.K. Heller, J.E. Ayers, and F.C. Jain, "Indium Gallium Arsenide Quantum Dot Gate Field-Effect Transistor Using II-VI Tunnel Insulators Showing Three-State Behavior", *J. Electronic Materials*, **41**, 10, pp. 2810- 2815 (2012).

16. P.-Y. Chan, M. Gogna, E. Suarez, F. Al-Amoody, S. Karmakar, B.I. Miller, E.K. Heller, J.E. Ayers, and F.C. Jain, "Fabrication and Simulation of an Indium Gallium Arsenide Quantum-Dot-Gate Field-Effect Transistor (QDG-FET) with ZnMgS as a Tunnel Gate Insulator", *J. Electronic Materials*, **42**, 11, pp. 3259-3266 (2013).
17. M. Lingalugari, P.-Y. Chan, E. Heller, and F.C. Jain, "Two-bit Quantum Dot Nonvolatile Memory (QDNVM) Using Cladded Germanium and Silicon Quantum Dots", *International Journal of High Speed Electronics and Systems*, **42**, 11, pp. 3275-3282 (2013).
18. V.A. Gritsenko, K.A. Nasyrov, Yu.N. Novikov, A.L. Aseev, S.Y. Yoon, Jo-Won Lee, E.-H. Lee, C.W. Kim, "A new low Voltage fast SONOS Memory with high-k Dielectric." *Solid-State Electronics*, **47**, pp. 1652 (2003).

Denoising and Beat Detection of ECG Signal by Using FPGA

Dheyaa Alhelal* and Miad Faezipour†

Department of Computer Science & Engineering and Biomedical Engineering,
Digital/Biomedical Embedded Systems and Technology Laboratory,
University of Bridgeport, Bridgeport, CT 06604, USA
*dalhelal@my.bridgeport.edu †mfaezipo@bridgeport.edu

This paper introduces an efficient digital system design using hardware concepts to filter the Electrocardiogram (ECG) signal and to detect QRS complex (beats). The system implementation has been done using a Field Programmable Gate Array (FPGA) in two phases. In the first phase, Finite Impulse Response (FIR) filters are designed for preprocessing and denoising the ECG signal. The filtered signal is then used as the input of the second phase to detect and classify the ECG beats. The entire system has been implemented on ALTERA DE II FPGA by desinging synthesizable finite state machines. The design has been tested on ECG waves from the MIT-BIH Arrhythmia database by windowing the signal and applying adaptive signal and noise thresholds in each window of processing. The hardware system has achieved an overall accuracy of 98% in the beat detection phase, while providing the detected beats and the classification of irregular heat-beat rates in real-time.

The synthesized hardware of the ECG denoising and beat detection system yields reasonable hardware resources, making the system attractive to be eventually fabricated as a stand alone hardware system or integrated/embedded within a portable electronic device for monitoring patients' heart conditions on a daily basis conveniently.

Keywords: beat detection; electrocardiogram (ECG); field programmable gate arrays (FPGA); finite impulse response (FIR) filter.

1. Introduction

The Electrocardiogram (ECG) is a medical diagnostic test which contains important information regarding the electrical activity of heart. The following represents information that supply from ECG signal pertaining to the human heart condition [1]:

- heart position and its relative chamber size
- impulse origin and propagation
- heart rhythm and conduction disturbances
- extent and location of myocardial ischemia
- changes in electrolyte concentrations
- drug effects on the heart

†Corresponding author.

One of the most conspicuous features in the electrocardiogram signal is the QRS complex. A QRS complex has a unique shape which provides important information regarding the current situation of the heart depending on its location (time) and amplitude of the ECG signal. Due to its discriminative features, it has become the basis for the determination and classification of the heart rate (arrhythmia) [2]. Generally, the ECG signal should be preprocessed and advanced signal processing techniques are applied to extract the important features of the ECG signal such as detecting the QRS complex. Preprocessing generally includes filtering the ECG signal to remove unwanted waves and sources of noise.

During the past decades, there have been many preprocessing and feature extraction algorithms performed on ECG signals and many of these algorithms have been developed and implemented. However, most of them deal with signal processing techniques that use software applications. Software applications have many advantages. For example, good accuracies can be achieved because the software process ideally runs on a virtual environment, and is often very easy to deal with. On the other hand, however, software applications do not carry out the results in real-time. ECG preprocessing in software requires saving the ECG signal, thus increasing the time/complexity and effort. Therefore, it is necessary to think about developing techniques that can be implemented on portable devices to perform the task of these software systems in real-time.

Here, we focus on the importance of digital systems, where features such as small size and efficiency make them attractive to accomplish what they were designed for. Recently, FPGA's captured significant attention compared to their hardware counterparts such as ASICs (Application Specific Integrated Circuit) and CPLDs (Complex Programmable Logic Device) because they are characterized by certain properties such as supporting rapid prototyping and performing custom processing at high data rates. Furthermore, FPGA allows for updating the system without the need to add or remove parts of the hardware. As a result, these capabilities provide a highly efficient upgradable system at a lower cost.

2. Related Work

There are many algorithms which have been proposed for ECG signal analysis and QRS detection. Some of them are derived from Artificial Neural Network [3, 4, 5, 6], Wavelet Transforms [7, 8, 9], Genetic Algorithms [10], and Filter Banks [11]. Most of ECG analysis and QRS detection techniques share the same algorithmic structure which is divided into two stages. The first stage is preprocessing stage, which includes filtering (Linear filtering) and feature extraction (Nonlinear filtering) while the second stage, which is the decision stage, includes peak detection and decision logic as shown in Fig. 1.

The authors in [12] are considered to be the pioneers who worked on the development of QRS detection algorithm. Their algorithm involved many steps of processing. The first step was passing the signal through Bandpass filter consisting of a cascade highpass and lowpass filters to reduce the effect of muscle noise, 60 Hz, T-wave interference and baseline wander. The next step was differentiation to provide the QRS slope information.

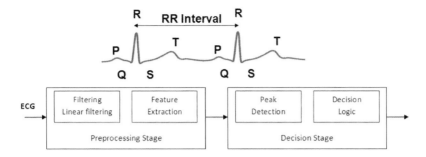

Fig. 1. Structure of the ECG QRS detectors.

In order to emphasize predominantly the ECG frequencies, they squared the signal point by point. At this stage the information regarding the slope and the width of the QRS complex are needed and could be obtained by the moving window Integrator. They deduced that the width of the QRS complex and the time of the rising edge are equal. As a result, a fiducial point could be determined from that rising edge. In their algorithm, they introduced two thresholds; the higher one was used for the initial analysis while the lower one was used when the search back technique is applied. Both of these thresholds were used when the QRS is not detected within a certain time interval. Determining the location (time-stamp) of the QRS complexes was the last step of their algorithm.

3. Proposed Methodology

As mentioned earlier, the ECG analysis and QRS detection algorithms in general share the same architecture represented in two stages. The preprocessing stage represents the first stage and peak detection represents the second stage as shown in Fig. 1. This section describes the ideas for hardware design/implementation of the above two stages.

3.1. Preprocessing Stage

The preprocessing stage involves many steps such as low pass filter, high pass filter, derivative filter, and moving-window Integration. This part of the hardware implementation is very important because it removes or at least reduces the entire unwanted signal (noise) such as baseline wander noise and power line interface noise which is added to the ECG signal. This prepares the signal for the next stage. The general view of the preprocessing stage is shown in Fig. 2.

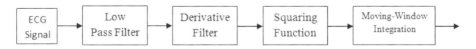

Fig. 2. General view of the preprocessing stage.

3.1.1. Low pass filter

In this part, Finite Impulse Response (FIR) filter will be implemented in hardware to reduce the powerline noise interference at 60 Hz frequencies and higher. However, there are two steps that must be taken into consideration before the digital FIR filter is implemented in hardware. First, specifications of the required filter in frequency domain should be known. Second, quantization of the filter coefficients must be performed to avoid the mismatch between the actual filter response and ideal filter response. The hardware design of the FIR filter can be obtained from implementing the equations (Eq. (1) and Eq. (2)) below:

$$Y(n) = \sum_{i=0}^{M} b(i) X(n-i) \qquad (1)$$

$$Y(n) = b(0)x(n) + b(1)x(n-1) + \cdots + b(M)X(n-M) \qquad (2)$$

Where:

$X[n]$ is the input signal
$Y[n]$ is the output signal
$b(i)$ are the filter coefficients (tap weights)
M is the filter order

Our filter will be a low pass filter with 360 Hz sampling frequency and the order equal to ten. That means the filter coefficients (tap) will be eleven. All the filter coefficients are eight bit signed integers as shown in Table 1:

Table 1. Filter Coefficients.

Coefficient	b0	b1	b2	b3	b4	b5	b6	b7	b8	b9	b10
8 bit Integers	-15	-13	7	38	66	78	66	38	7	-13	-15
Hexadecimal	F1	F3	07	26	42	4E	42	26	07	F3	F1

There are many structures for FIR filters. The direct form structure has been chosen to implement the FIR filter in hardware (FPGA), as shown in Fig. 3.

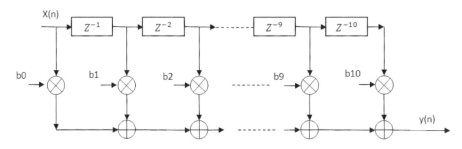

Fig. 3. Direct form of FIR filter.

3.1.2. Derivative filter

The Differentiator is used to provide the slope information of the QRS complex. This filter is widely used among the ECG analysis algorithms to find the slope information because the slope of the QRS complex is generally larger than the other features in the ECG waveform. The difference equation for the derivative filter is shown in the equation below [13]:

$$Y(nT) = (2x(nT) + x(nT - T) - x(nT - 3T) - 2x)nT - 4T))/8 \qquad (3)$$

3.1.3. Squaring function

The output of the previous stage which is a derivative filter will be the input of this stage. Squaring the ECG signal implies that the high frequencies will be emphasized and the entire data signal will be positive. The equation of the squaring function is shown in the equation below:

$$y(nT) = [x(nT)]^2 \qquad (4)$$

The moving window integration is a type of FIR filter that adds the N previous samples from the previous stage (Squaring Function) and then divides the sum by N. As a result, the output of the moving window integration will be the average of the N samples. The moving window integration is obtained from the equation below [12]:

$$y(nT) = \left(\frac{1}{N}\right)[x(nT - (N-1)T) + x(nT - (N-2)T) + \cdots + x(nT)] \qquad (5)$$

Generally, the width of the moving window integration (N) and the QRS complex should be the same as much as possible. The reason behind this is that the QRS and T complexes will be merged if the width of the moving window integration is greater than QRS complex, while there are several peaks in the integration waveform that will be produced by some QRS complexes if the width of the moving window integration is less than QRS complex. As a result, the number of samples (N) in the moving window integration is determined empirically [12].

3.2. Beat Detection Stage

Many algorithms have been introduced for ECG beat detection. The authors in [14], [15] proposed modification to the existing Pan and Tompkins algorithm where they reduced the data processing by introducing an algorithm which has only one set of adaptive threshold computations instead of two in Pan and Tompkins algorithm [12]. The algorithm in [14], [15] is adopted in the hardware implementation of our system because it has less data processing. Consequently, the digital logic components of our system will be less. The equations for adaptive thresholding are as below:

$$\text{SPK} = 0.125 \times \text{PEAK} + 0.875 \times \text{SPK} \qquad (6)$$

where PEAK = Maximum (Peaks), and

$$THR = NPK + 0.25 (SPK - NPK) \tag{7}$$

where NPK = Minimum (Peaks)

After that, each signal peak will be compared with the threshold THR. If the signal peak is greater than the threshold, that signal will be considered as a QRS complex. According to the original Pan and Tompkins algorithm, the next R point can not be detected in 360 ms from the previous R point. Therefore, whenever an R point is found, the next start point will be after 360 ms. However, if the R point (beat) is not detected within a certain time (116% of the current R to R average), a search back technique will be applied. Another data processing reduction that has been applied in this algorithm is computing one R point to R point average instead of two in Pan and Tompkins algorithm. The equation for R to R average is shown below.

$$RR_{Avg} = \frac{1}{n-1} \sum_{i=0}^{n-2} (RR_{(n-i)}) \quad 2 \le n \le 7 \tag{8}$$

$$RR_{Avg} = \frac{1}{8} \sum_{i=0}^{7} (RR_{(n-i)}) \quad n \ge 8 \tag{9}$$

It is important to note that the threshold will be different when the search back is applied. The equation for the threshold when the search-back is applied is shown below:

$$THR_{new} = \frac{THR_{old}}{8} \tag{10}$$

In order to implement this algorithm in hardware, two separate processes are performed. The first process is to store the incoming ECG signal samples in memory registers and at the same time find the maximum and minimum peaks to be able to start with the second process which is a finite state machine (FSM). This FSM controls the beat detection process and consists of nine states as shown in Fig. 4.

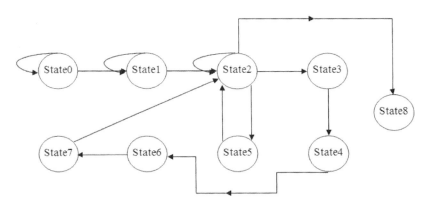

Fig. 4. State diagram of the FSM.

The nine states are described hereafter:

State0

This state is the initial state. The system will stay in this state until the initial signal which is coming from the previous process becomes 1. The next state will be state1.

State1

The system remains in State1 while determining the first threshold. The first 360 samples from the ECG signal are used to calculate the first threshold. The next state will be state2.

State2

State2 is the brain of the system. There are many conditions which determine the next state. The first condition is the main condition where it checks if the sample is the last sample or not. If the sample is the last sample in the ECG signal, the next state will be state8 where the final result will be displayed. Otherwise, the system will be checking the next condition where the peak signal will be compared against the threshold. If it is larger than the threshold, the next state will be state3, otherwise the system will compare the number of samples from the previous beat detected with 116% of the current RR average. If the number of samples is greater and no beat is detected, the next state will be state5 where the search back algorithm will be applied using a new threshold. If the previous two conditions were false, the next state will be the same state.

State3

Let us call this state the beat detection state. Since the beat is detected, the threshold should be recalculated. After that, the next state will be state4.

State4

In this state, the last eight RR intervals will be stored in a register by using shift registers, and then the average will be calculated. After that, the next state will be state6.

State5

As mentioned before, the threshold will be calculated and the next state will be state2.

State6

In state6 all the data will be available. Therefore, the condition which will determine if the search back algorithm will be applied or not, will be evaluated here. Since all values are available at this state, the heart rate is calculated here ($RR_{interval}$). After that, the next state will be state7.

State7

The heart rate is calculated in the previous state. Therefore, a classification procedure is then applied to check if the heart rate falls in the normal range or not (92%-116% of the average RR). In order to repeat all these processes for all samples signals, the next state for state7 is state2.

State8

In this state the final output will be displayed.

4. Results

The FSM described in Section 3 has been implemented in the ALTERA DE II FPGA device. The hardware resources are summarized in Table 2.

Table 2. Resource utilization of the hardware.

Total logic elements	13,777 out of 18,752	73%
Total combinational functions	13,611 out of 18,752	73%
Dedicated logic registers	3,532 out of 18,752	19%
Total registers	3,532	
Total pins	128 out of 152	84%
Total memory bits	131,072 out of 239,616	55%
Embedded Multiplier 9-bit elements	20 out of 52	38%

The data records used to test the design were obtained from the MIT-BIH Arrhythmia database from the Physionet website [16]. The data records were sampled at 360 Hz. As mentioned before, the QRS complex detection part is divided into two stages; preprocessing stage and peak detection stage. Therefore, the results will follow the same division. Figure 5 shows the resulted waveforms in the preprocessing stage for one of the ECG records of the dataset.

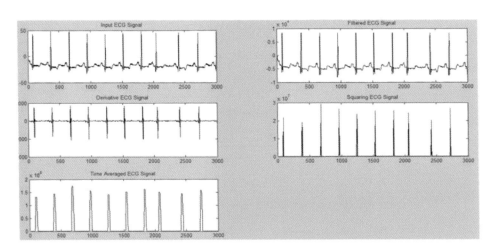

Fig. 5. Preprocessing stage waveform results.

The digital system is tested using Quartus II tool and the beat detection performance for 5 of the data records are shown in Table 3. The overall accuracy of the entire dataset yielded a result of 98%, which is comparable with software based signal processing techniques, while producing the results in real-time.

Table 3. Performance Evaluation of Beat Detection.

Data #	# of Beats	Not Detected	Error
100	73	1	1.37%
101	67	4	5.97%
102	72	1	1.39%
103	69	1	1.45%
104	74	0	0%
105	82	1	1.22%
Total	437	8	Overall Error 1.9%

5. Comparison with other Similar Work

Most ECG denoising and beat detection approaches are evaluated to meet the requirements for real-time performance and power efficiency as well as complexity to be suitable for hardware implementations [17]. Some hardware techniques are based are based on Digital Signal Processing (DSP) processors. FPGA digital hardware approaches, on the other hand, offer an advantage over DSP processors as they exceed the computing power by breaking the paradigm of sequential execution and accomplishing more operations per clock cycle.

Table 4 provides a list of a few recent hardware based approaches and their performances in comparison with our proposed work.

Table 4. Comparison of a few recent ECG analysis systems.

Reference/ Year	Hardware/Software Platform	ECG Dataset	Efficiency for Real-time Hardware	Performance
[17]/2014	Mobile Phone Software Applications	PhysioNet's MIT-BIH Arrhythmia Database	Medium	97%-99% Sensitivity and Positive Predictivity Rates
[18]/2016	Xilinx FPGA	PhysioNet's MIT-BIH Arrhythmia Database	High	97.8% Accuracy
[19]/2016	TMS320C6713 DSP	PhysioNet's MIT-BIH Arrhythmia Database	High	98.89% Accuracy for ECG Denoising only
[20]/2016	Altera FPGA	PhysioNet's MIT-BIH Arrhythmia Database	High	N/A Time Complexity Most Efficient for MAC Pipelined Implementation
Proposed	Altera FPGA	PhysioNet's MIT-BIH Arrhythmia Database	High	98% Accuracy

6. Conclusion

The main goal of this work was to design an efficient digital system to filter and analyze the ECG signal in real-time. The FIR filter was implemented in hardware as the first phase of the system and an efficient beat detection algorithm was implemented in hardware as the second phase. The entire system was designed by using FPGA (ALTERA DE II) where Quartus II toolset was used to simulate and test the system. This framework is especially useful for real-time ECG signal analysis for patients with severe heart diseases.

The synthesized hardware of the ECG denoising and beat detection system yields reasonable hardware resources such as logic elements of the FPGA. Therefore, the proposed digital system framework can eventually be fabricated as a stand-alone hardware system or integrated/embedded within a portable electronic device such as a smart-phone for convenient daily basis ECG monitoring.

References

1. J. Sahoo, "Analysis of ECG signal for Detection of Cardiac Arrhythmias", *Master's Thesis in Telematics and Signal Processing, Department of Electronics and Communication Engineering, National Institute Of Technology, India*, 2011.
2. B.-U. Kohler, K. Hennig, and R. Orgelmister, "The Principles of Software QRS Detection" *IEEE Engineering In Medicine And Biology Magazine,* vol. 21, no. 2, pp. 42-57, 2002.
3. Y. H. Hu, W. J. Tompkins, J. L. Urrusti, and V. X. Afonso, "Applications of artificial neural-networks for ECG signal detection and classification," *J. Electrocardiology*, vol. 26 (Suppl.), pp. 66-73, 1993.
4. M. G. Strintzis, G. Stalidis, X. Magnisalis, and N. Maglaveras, "Use of neural networks for electrocardiogram (ECG) feature extraction, recognition and classification," *Neural Netw. World*, vol. 3, no. 4, pp. 313-327, 1992.
5. G. Vijaya, V. Kumar, and H. K. Verma, "ANN-based QRS-complex analysis of ECG," *J. Med. Eng. Technol.*, vol. 22, no. 4, pp. 160-167, 1998.
6. Q. Xue, Y. H. Hu, and W. J. Tompkins, "Neural-network-based adaptive matched filtering for QRS detection," *IEEE Trans. Biomed.Eng.*, vol. 39, pp. 317-329, 1992.
7. S. Kadambe, R. Murray, and G. F. Boudreaux-Bartels, "Wavelet transform-based QRS complex detector," *IEEE Trans. Biomed. Eng.*, vol. 46, pp. 838-848, 1999.
8. C. Li, C. Zheng, and C. Tai, "Detection of ECG characteristic points using wavelet transforms," *IEEE Trans. Biomed. Eng.*, vol. 42, no. 1, pp. 21-28, 1995.
9. K. D. Rao, "DWT based detection of R-peaks and data compression of ECG signals," *IETE J. Res.*, vol. 43, no. 5, pp. 345-349, 1997.
10. R. Poli, S. Cagnoni, and G. Valli, "Genetic design of optimum linear and nonlinear QRS detectors," *IEEE Trans. Biomed. Eng.*, vol. 42, pp. 1137-1141, 1995.
11. V. X. Afonso, W. J. Tompkins, T. Q. Nguyen, and S. Luo, "ECG beat detection using filter banks," *IEEE Trans. Biomed. Eng.*, vol. 46, pp. 192-202, 1999.
12. J. Pan and W. J. Tompkins, "A real-time QRS detection algorithm," *IEEE Trans. Biomed. Eng.*, vol. BME-32, no. 3, pp. 230-236, Mar. 1985.
13. Patrick S. Hamilton and Willis J. Tompkins, "Quantitative Investigation of QRS Detection Rules Using the MIT/BIH Arrhythmia Database", *IEEE Transactions On Biomedical Engineering*, vol. BME-33, no. 12, pp. 1157-1165, Dec. 1986.
14. M. Faezipour, T. M. Tiwari, A. Saeed, M. Nourani, and L. S. Tamil, "Wavelet-Based Denoising and Beat Detection of ECG Signal," in *IEEE/NIH Life Science Systems and Applications Workshop*, pp. 100-103 (LiSSA 2009).

15. M. Faezipour, A. Saeed, S. C. Bulusu, M. Nourani, H. Minn and L. S. Tamil, "A Patient-Adaptive Profiling Scheme for ECG Beat Classification", in *IEEE Transactions on Information Technology in Biomedicine*, vol. 14, no. 5, pp. 1153-1165, Sep. 2010.
16. PhysioNet: *www.physionet.org*.
17. M. Elgendi, B. Eskofier, S. Dokos and D. Abbott, "Revisiting QRS Detection Methodologies for Portable, Wearable, Battery-Operated, and Wireless ECG Systems", in *PLoS One Journal*, vol. 9, no. 1, e84018, 2014.
18. M. G. Egilaa, M. A. El-Moursyb, A. E. El-Hennawyc, H. A. El-Simarya and A. Zakia, "FPGA-based electrocardiography (ECG) signal analysis system using least-square linear phase finite impulse response (FIR) filter", in *Journal of Electrical Systems and Information Technology*, vol. 3, no. 3, pp. 513-526, Dec. 2016.
19. S. A. Anapagamini and R. Rajavel, "Hardware implementation of ECG denoising system using TMS320C6713 DSP processor", in *International Journal of Biomedical Engineering and Technology*, vol. 21, no. 1, pp. 95-108, 2016.
20. S. Al-Khammasi, K. A. I. Aboalayon, M. Daneshzand, M. Faezipour and M. Faezipour, "Hardware-Based FIR Filter Implementations for ECG Signal Denoising: A Monitoring Framework from Industrial Electronics Perspective", in *Proceedings of the Annual IEEE Connecticut Conference on Industrial Electronics, Technology & Automation*, pp. 1-6, Oct. 2016.

Encrypted Electron Beam Lithography Nano-Signatures for Authentication

Kiarash Ahi[*], Abdiel Rivera and Mehdi Anwar

Department of Electrical Engineering,
University of Connecticut,
Storrs, CT 06268, USA
[*]*kiarash.ahi@uconn.edu*

In this work, engineered nanostructures (ENS) have been fabricated on the packed integrated circuits. Coding lookup tables were developed to assign different digits in numerical matrices to different fabricated nano-signatures. The numerical matrices are encrypted according to advanced encryption standard (AES). The encrypted numerical matrix is ink printed on the components, and the nano-signatures are fabricated on the packaged of the chips via electron beam lithography (EBL). This process is to be done in the manufacturer side of the supply chain. The numerical matrix and the nano-signature accompany the product in its long journey in the global supply chain. The global supply chain is proved to be susceptible to counterfeiters. For keeping counterfeiters' hands out of the process, the cipher key and the coding lookup tables are provided to the consumer using a secure direct line between the authentic manufacturer and the consumer. In the consumer side, the printed numerical matrix is decrypted. Having the decrypted numerical matrix makes it possible to extract the nano-signature from the laser speckle pattern shined on the packaged product. In this work, an algorithm is developed to extract the nano-signature by having the decrypted matrix and reflected laser speckle patterns as inputs. Confirming the existence of the nano-signature confirms the authenticity of the component. Imitating the nano-signatures by the counterfeiters is not possible because there is no way for them to observe the shape of these signatures without having access to the cipher key.

Keywords: electron beam lithography; counterfeit detection; engineered nanostructures; image processing; SSIM; AES.

1. Introduction

In this work, engineered nanostructures (ENS) have been fabricated on the surfaces of integrated circuits. A method using image processing and pattern recognition was developed to read the physical patterns of the nano-signatures. Encryption methods are adopted to keep the nano-signatures hidden from the counterfeiters.

The financial loss of supply chains due to counterfeit electronic components is reported to be higher than $10 billion per year [1]. This loss would be much more tremendous by considering the equipment failures because of malfunctions of counterfeit electronic components. The loss of lives due to these failures is beyond financial measures. In a more general view, the counterfeiting and pirated products have imposed catastrophic damages

---

[*]Corresponding author.

on the economies worldwide. The total value of counterfeit and pirated products in for G20 countries are reported to be up to $650 billion in 2008 and it is predicted to rise to $1,770 billion in 2015 [2]. Overdesigning the systems by adding modular redundancies adds to the costs, volumes and weights of the systems, which are crucial factors for critical systems such as aerospace and transportation systems, and thus it is preferred to enhance the reliability of a system by using reliable components other than adding redundancies [3], [4]. As the electronic industry is growing fast, counterfeit electronic components bring more revenue to the counterfeiters and thus, more electronic counterfeit components are injected into the market. The counterfeit electronic component incident reports have closely followed the global semiconductor revenue so far. Consequently, since a sharp rise in electronic components market is predicted, a high rise in counterfeit components is expected. Most of the methods for detection of counterfeit ICs have been based on physical inspections and electric tests [5]–[13]. As counterfeiters are making their approaches more advanced, the need for adding signatures on the electronic components is arising [14]. Such signatures do not only need to be invisible from counterfeiters but also not to be replicable by counterfeiters. These signatures are proposed as essential authentication means for industrial and healthcare electronics components [15]–[17].

In this work, complex matrix patterns were realized by putting different sizes of nanostructures as entries in a matrix. These signature matrices which entries are ENS were formed to embed passcodes on the surfaces of the ICs. A novel method is developed for translating random patterns of ENS to the encrypted numerical matrices. The numerical matrices are encrypted using advanced encryption standard (AES) then the encrypted matrices are translated to the ENS matrices patterns using the developed ENS lookup table. ENS matrices are not visible unless the observer knows the assigned patterns in advanced. Image matching between the translated ENS pattern from decryption of numerical matrix and the ENS pattern extracted from speckle laser pattern of the surface of the object proves the authentication of the object. The encrypted numerical matrix is printed in ink on the packages of the objects and is obvious to everyone. The ENS pattern, however, is fabricated using electron beam lithography (EBL) and is invisible.

The numerical matrix cannot be translated to the ENS pattern unless the cipher key and the dedicated ENS lookup table are both known. The fabricated ENS matrix cannot be extracted unless its pattern is known from the decrypted numerical matrix. Thus, counterfeiters cannot extract the ENS pattern shape from the surface of the object while customers can obtain the key from the authentic manufacturer. In this way, counterfeiters cannot replicate the ENS by reversed engineering. The flexibility of making random signatures, encryption with AES together with the possibility of dedicating a unique ENS lookup table for each series of the product makes it impossible for the counterfeiters to broke into the process and replicate it.

This paper is organized as follows. In Section II, the concepts of the nano-signatures, encryption and physical fabrication via EBL are discussed. In Section III, the image processing approach for extraction of the ENS patterns from the laser speckle patterns based on matching concepts is developed. In Section IV, the theory is implemented in practice and is confirmed by practical results.

2. Concepts, encryption, and fabrication of the nano-signatures

The nano-signature matrix which is comprised of ENS entries is fabricated on the surface of the IC. Metamaterials with negative refractive indices are used for fabrication of the ENS matrix. The fabrication process and reliability of the process is discussed in [15], [18]–[20]. This matrix is too small to be seen by an optical microscope and since the surface of the IC is not conductive, it is invisible to electron microscopes. Thus, counterfeiters cannot access to the shape of the matrix to replicate it. An algorithm is developed to extract the pattern of this matrix from the speckle pattern. This algorithm is discussed in the next section. However, for extraction of this matrix from the speckle pattern, the ENS pattern should be known in advance. In fact, the algorithm matches the known ENS pattern with the speckle pattern. If a match occurs, the authentication is done. Otherwise, the object is distinguished to be counterfeit. The whole process of the authentication is depicted in Fig. 1. The ENS is fabricated by the manufacturer on the surface of the packaged ICs. A numerical matrix is also printed on the package of the IC. This numerical matrix which is vivid to everyone is encrypted using advanced encryption standard (AES) [21]. A lookup table is also provided to convert the decrypted numerical matrix to the ENS pattern. Once this matrix is decrypted and converted to the ENS matrix pattern, authentication of the product is proved. An example of the ENS lookup table is brought by Table-I. An ENS element and the conversion (coding) of the numerical matrix to the ENS pattern together with the AES encryption of the numerical matrix are shown in Figs. 2 and 3 respectively.

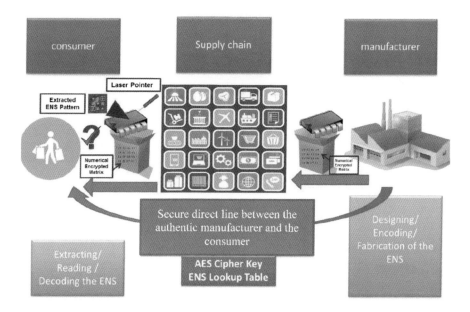

Fig. 1. The process of identification of authentic components by nano-signatures.

Table-I. An example of the ENS lookup table.

ENS Location \ Sizes	5 µm	10µm	15 µm	20 µm
1x1	32	55	3e	ec
1x2	80	a8	90	e5
1x3	79	6c	31	82
1x4	6b	7f	4f	e0
2x1	c8	12	43	06
2x2	5a	8e	2b	f3
...	...	...	...	...
4x4	34	e6	52	4d

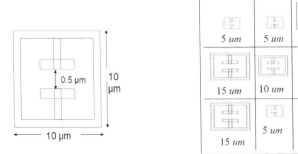

Fig. 2. (a) A 10 µm ENS. (b) A random 3 × 3 matrix of ENS.

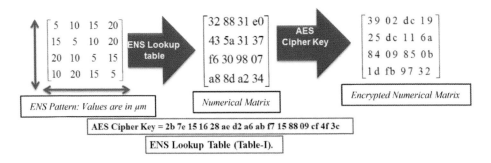

Fig. 3. Left to right: Converting an ENS pattern to a numerical matrix using the ENS lookup table (Table-I), and then encrypting the numerical matrix using AES. The right-hand matrix is ink printed on the product while the left-hand side matrix is converted to a signature-shaped pattern and fabricated via EBL.

3. Extraction of the fabricated ENS to match the signature matrix

The ENS array can be tailored to provide unique signatures for each and every piece of product. This random signature either unique to a family if ICs or an individual IC may be entered into an IC identifier register for future reference and cross-verification. This will allow the detection of over-produced or counterfeit ICs as the counterfeiters will not be able to regenerate the random ENS unique to an individual IC even if they are able to imprint one particular ENS. Once the numerical matrix is converted to the ENS pattern matrix using the AES cipher key and ENS lookup table, in a reversed process of that of the Fig. 3, the developed algorithm in this section can match the fabricated pattern with the decrypted matrix to prove the authentication. Without having the knowledge of the ENS pattern this algorithm cannot adjust the similarity window precisely and thus even if the counterfeiters break into this algorithm, they cannot extract the fabricated ENS without having access to the ENS lookup table and the cipher key.

For extraction of the fabricated ENS from the laser speckle pattern, the SSIM algorithm is modified as follows [22]. SSIM takes three parameters into account, namely, luminance, contrast, and structure. The algorithm takes a window of $n \times m$ pixels, which upper left corner located at $x \times y$, and calculates the $SSIM(x, y)$ for that window. Consequently, when two $w \times l$-pixel images are given to the SSIM algorithm, $(w-n+1) \times (l-m+1)$ values are calculated as $SSIM(x, y)$ for each pixel pairs. Eventually, the algorithm gives the average of all the calculated $SSIM(x, y)$ as the SSIM index of the two images.

For detecting ENS in an obtained image from laser reflection from the surface of an IC, SSIM index can be employed. In this application, in $SSIM(x, y)$ formula which is represented by Eq. (1), luminance, $l(x, y)$ would not be effective in the detection of the ENS since in general it is desired to have a luminance independent detection system. From formulation point of view, luminance is dependent only on the mean values of the pixel sets which occurs inside the mentioned window; since the mean value would be highly determined by the background noise in case of the images of this work, it`s desired to eliminate the effect of the luminance in the overall calculations. Contrast, $c(x, y)$ is not as important as the structure $s(x, y)$ factor as well. So α which indicates weight of luminance is set to zero for this work, and β, the weight of contrast is set much lower than γ the weight of $c(x, y)$. Then, the resulting formula for detecting ENS would be equal to $s(x, y)$ which is indicated by Eq. (1).

$$SSIM(x, y) = s(x, y) = \frac{\sigma_{xy} + C}{\sigma_x \sigma_y + C} \quad (1)$$

Where σ_x and σ_y are the standard deviations of the two-pixel sets held inside the mentioned window from the two images, one image is indicated by x and the other one by y. C is a constant and σ_{xy} is the sample cross-correlation of x and y after removing their means [23], [24].

$$\sigma_{xy} = \frac{1}{N-1} \sum_{i=1}^{N} (x_i - \mu_x)(y_i - \mu_y) \quad (2)$$

3.1. Detecting the location of the ENS as the entries of the signature matrix

The sizes and locations of the fine structural similarities caused by the reflection from ENS are not known within the image of the reflected laser beam from the surface of the IC. To detect the sizes and locations of the reflections of the ENS within the obtained images, the $SSIM(x,y)$ which is indicated in Eq. (1) is calculated inside the mentioned selection window. In each iteration, the size of the window varies from 2×2 pixels to the size of the image. For calculating the $SSIM(x,y)$ over the entire image, location of the windows moves from $(x,y)=(0,0)$ to $(x,y)=(w-n+1)\times(l-m+1)$. In another word, the $SSIM(x,y)$ is calculated for $(x,y)=(w-n+1),(l-m+1)$ in the windows of the sizes 2×2 to $w\times l$ where $0\leq x\leq w-n+1$ and $0\leq y\leq l-m+1$. Once the window covers a similar structure inside the two images, say at the location $(x,y)=(x_1,y_1)$, a higher calculated $SSIM(x_1,y_1)$ is observed compared to $SSIM(x,y)$ in the neighborhood of (x_1,y_1). Consequently, once a peak is observed in the results, the location of the corresponding ENS is detected to be $(x_1+\frac{n}{2}, y_1+\frac{m}{2})$.

3.2. Detection of the size of the ENS

For obtaining the size of the reflected ENS it should be taken into account that as the window reaches the ENS, a peak is going to be shaped in the resulting set of SSIM as it is illustrated in Fig. 4. Where the size of the window is larger than the ENS reflection, the peak will not be sharp, because the window includes the ENS reflection during several iterative movements. Where the window is smaller, sharp peaks will not be observed as the window cannot cover the whole ENS reflection and in another word the ENS occurs inside the window during several iterative movements. Consequently, where a well-shaped peak is observed, the size of the respective window refers to the size of the reflection of the ENS. One can consider the moving window as a square signal which is convolved to another square signal (which is the ENS reflection), thus in practice, half of the width of the peak represents the size of the ENS.

4. Practical results

A pattern consisting of different sizes of ENS is written on the surface of an IC. This pattern is shown in Fig. 5. Each element in the matrix refers to the size of each of the ENS. For example, the second diagonal element refers to a $5\mu m \times 5\mu m$ ENS which is surrounded by eight other ENS neighbors with sizes $15\mu m\times 15\mu m$, $5\mu m\times 5\mu m$, $10\mu m\times 10\mu m$, $15\mu m\times 15\mu m$, $10\mu m\times 10\mu m$, $5\mu m\times 5\mu m$, $10\mu m\times 10\mu m$ and $20\mu m\times 20\mu m$ clockwise. The IC is setup horizontally and thus the ratio (horizontal/vertical) of dimensions in this work is 0.63.

The speckle pattern shown in Fig. 6 is obtained from reflection of the laser beam from the surface of the IC. The aspect ratio of the speckle pattern used in the calculation of this section is 531×841-pixel.

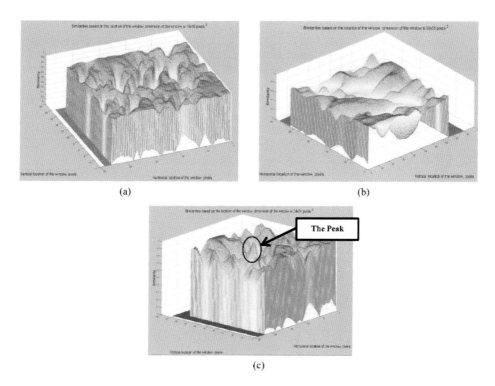

Fig. 4. (a) By applying windows of the dimensions less than 24×24 pixels almost no local peak is distinguishable in the mesh grids. (b) By applying windows of the dimensions larger than 24×24 pixels almost all the local peaks are smoothed in the mesh grid. (c) By applying windows of the dimensions 24×24 pixels local peaks are distinguishable in the mesh grids; this refers to the fact that the window fit the fine structures.

$$\begin{bmatrix} 5 & 10 & 15 & 20 \\ 15 & 5 & 10 & 20 \\ 20 & 10 & 5 & 15 \\ 10 & 20 & 15 & 5 \end{bmatrix}$$

Fig. 5. The fabricated ENS pattern: the entries of the matrix indicate the sizes of the ENS cells in μm.

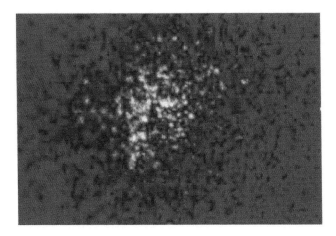

Fig. 6. The speckle pattern from reflection of the laser beam from the surface of the IC.

For detecting ENS, the method described in Section IV is applied to the speckle pattern. The speckle pattern is rotated 180 degrees for obtaining the second image to be employed by the algorithm. It is observed that as the size of the comparison window increases, some peaks appear. The peaks become sharper and sharper until the window reaches some certain size then again the peaks become smoother and smoother as the size of the window increases until the new sets of peaks appear in higher sizes. In this work, the first set of peaks appears when the size of the window reaches 50×50 pixels as it is shown in Fig. 7(a). According to Section IV, it is possible to obtain the size of the ENS by measuring the size of the corresponding peak. Towards this aim, the size of the peak is measured in the mesh grid of Fig. 7(a) perpendicular to the xy plane as it is shown in Fig. 7(b). The horizontal location of the peak is placed at pixel 593 and its hillside continues to 593 as it is indicated by the black arrow on Fig. 7(b), thus the length of the horizontal hillside is $(593 - 563) = 30\, pixels$. Similarly, the total length of the vertical hillside of this peak is $100\, pixels$. By incorporating the aspect ratio, the size of the corresponding ENS is obtaining as follows.

$$Vertical = (100 / 2\, pixels) \times 84um / (840\, pixels) = 5\ um$$
$$Horizontal = (30\ pixels) \times 53um / (531\, pixels \times 0.63) = 4.8\ um \tag{3}$$

As the size of window increases from 50×50 the mentioned set of peaks become smoother and then again a new set of peaks appears at 100×100, as shown in Fig. 4(c). By the same procedures and calculations, as presented in Eq. (4), it is revealed that these peaks are corresponding to ENS which size is $10\mu m$.

$$Vertical = (100\, pixels) \times 84um / (840\, pixels) = 10\ um$$
$$Horizontal = (63\ pixels) \times 53um / (531\, pixels \times 0.63) = 10\ um \tag{4}$$

Fig. 7. (a) The appeared peaks where the size of the window reaches 50 × 50 pixel. (b) The horizontal location of the peaks for the window 50 × 50 pixel. (c) The appeared peaks where the size of the window reaches 100 × 100 pixel. (d) The horizontal location of the peaks for the window 100 × 100 pixel. (e) The appeared peaks where the size of the window reaches 150 × 150 pixel. (f) The horizontal location of the peaks for the 150 × 150 pixel window. (g) The appeared peaks where the size of the window reaches 200 × 200 pixel. (h) The horizontal location of the peaks for the 200 × 200 pixel window.

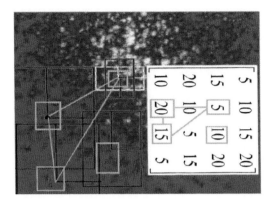

Fig. 8. Collection of the detected peaks on the same plane (the original image).

As the size of window increases, two other set of sharp peaks appears where the window is 150×150 and 200×200 pixels. By The same calculations as Eq. (4), it is obtained that these peaks are revealing the 15μm and 20μm ENS respectively.

Collecting the locations of all of the detected peaks on the same plane results in Fig. 8. Figure A in the Appendix I illustrates the detailed process of collecting the locations of the peaks. It was expected that the only symmetric elements of the matrix of Fig. 2 are detectable since the 180 degrees' rotation of the main image is employed as the second image for the procedure. As Fig. 5 indicates, only symmetrical elements of the matrix are revealed by this procedure. It is also observed that the detected locations are inconsistent with the location of the fabricated ENS.

5. Conclusion

In this paper, an innovative method for authentication of the electronic components has been represented. Engineered nano-signatures have been used for this purpose. These signatures are invisible to counterfeiters, and thus counterfeiters cannot clone them. In addition, since these signatures are fabricated on the packages of the components, sanding and blacktopping of the components by counterfeiters destroy them. The nano-signatures are encoded to numerical matrices by developing lookup tables. The numerical matrices are then encrypted using the advanced encryption standard algorithm. An image processing algorithm has been developed to extract the nano-signature from the reflected laser speckle pattern. This algorithm confirms if the correct nano-signature, which matches the decrypted numerical matrices, exists on the object. The numerical matrix needs to be decrypted before giving to the matching algorithm. The cipher key is transferred via a secure direct line between the consumer and the manufacturer. Consequently, counterfeit components which might have been injected to the supply chain are distinguished at the consumer end.

Appendix I

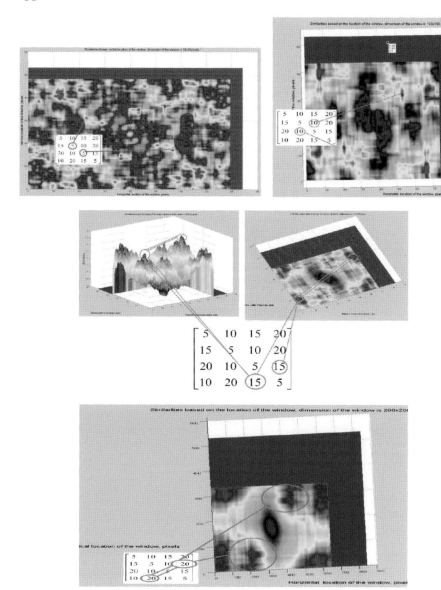

Fig. A. Assigning the detected peaks to the fabricated ENS pattern.

References

[1] J. M. Radman and D. D. Phillips, "Novel Approaches for the Detection of Counterfeit Electronic Components," *IN Compliance Magazine*, no. October, 2010.

[2] Frontier Economics Ltd., "Estimating the global economic and social impact of counterfeiting and piracy," 2011.

[3] K. Ahi, "Control and Design of a Modular Converter System for Vehicle Applications," Leibniz Universität Hannover, Hannover-Germany, 2012.

[4] F. A. Tillman, C. L. Hwang, and W. Kuo, "Determining Component Reliability and Redundancy for Optimum System Reliability," *IEEE Trans. Reliab.*, vol. R-26, no. 3, pp. 162–165, 1977.

[5] K. Ahi and M. Anwar, "Modeling of terahertz images based on x-ray images: a novel approach for verification of terahertz images and identification of objects with fine details beyond terahertz resolution," in *Proc. SPIE 9856, Terahertz Physics, Devices, and Systems X: Advanced Applications in Industry and Defense, 985610*, 2016, p. 985610.

[6] K. Ahi and M. Anwar, "Advanced terahertz techniques for quality control and counterfeit detection," in *Proc. SPIE 9856, Terahertz Physics, Devices, and Systems X: Advanced Applications in Industry and Defense, 98560G*, 2016, p. 98560G.

[7] A. Rivera, A. Mazady, K. Ahi, and M. Anwar, "Growth dependent optical properties of ZnMgO at THz frequencies," in *Proc. SPIE 9483, Terahertz Physics, Devices, and Systems IX: Advanced Applications in Industry and Defense, 94830X*, 2015, vol. 9483, p. 94830X.

[8] K. Ahi and M. Anwar, "Developing terahertz imaging equation and enhancement of the resolution of terahertz images using deconvolution," in *Proc. SPIE 9856, Terahertz Physics, Devices, and Systems X: Advanced Applications in Industry and Defense, 98560N*, 2016, p. 98560N.

[9] K. Ahi, N. Asadizanjani, S. Shahbazmohamadi, M. Tehranipoor, and M. Anwar, "Terahertz characterization of electronic components and comparison of terahertz imaging with x-ray imaging techniques," in *Proc. SPIE 9483, Terahertz Physics, Devices, and Systems IX: Advanced Applications in Industry and Defense, 94830K*, 2015, p. 94830K.

[10] K. Ahi and M. Anwar, "A Novel Approach for Enhancement of the Resolution of Terahertz Measurements for Quality Control and Counterfeit Detection," in *Diminishing Manufacturing Sources and Material Shortages (DMSMS)*, 2015.

[11] IHS, "Counterfeit-Part Risk Expected to Rise as Semiconductor Market Shifts into Higher Gear," EL SEGUNDO, Calif., 2012.

[12] K. Ahi, N. Asadizanjani, M. Tehranipoor, and M. Anwar, "Authentication of electronic components by time domain THz Techniques," in *Connecticut Symposium on Microelectronics & Optoelectronics*, 2015.

[13] K. Ahi, N. Asadizanjani, S. Shahbazmohamadi, M. Tehranipoor, and M. Anwar, "THZ Techniques: A Promising Platform for Authentication of Electronic Components," in *CHASE Conference on Trustworthy Systems and Supply Chain Assurance*, 2015.

[14] Z. Ali and B. Florent, "Potential of Chipless Authentication Based on Randomness Inherent in Fabrication Process for RF and THz," in *Conference: 11th European Conference on Antennas and Propagation*, 2017.

[15] K. Ahi, A. Rivera, A. Mazady, and M. Anwar, "Authentication of electronic components using embedded nano-signatures," in *Connecticut Symposium on Microelectronics & Optoelectronics*, 2015.

[16] K. Ahi and M. Anwar, "Embedding and Fabrication of Authentication Signatures by Robust Engineered Nanostructures," in *Diminishing Manufacturing Sources and Material Shortages (DMSMS)*, 2015.

[17] K. Ahi and M. Anwar, "A Novel Approach for Extracting Embedded Authentication Engineered NanoSignatures," in *Diminishing Manufacturing Sources and Material Shortages (DMSMS)*.
[18] K. Ahi, A. Mazady, A. Rivera, M. Tehranipoor, and M. Anwar, "Multi-level Authentication Platform Using Electronic Nano-Signatures," in *2nd International Conference and Exhibition on Lasers, Optics & Photonics*, 2014.
[19] K. Ahi, A. Rivera, A. Mazady, and M. Anwar, "Fabrication of robust nano-signatures for identification of authentic electronic components and counterfeit avoidance," *Int. J. High Speed Electron. Syst.*, 2017.
[20] K. Ahi, A. Rivera, A. Mazady, and M. Anwar, "Embedding Complex Nano-Signatures for Counterfeit Prevention in Electronic Components," in *CHASE Conference on Trustworthy Systems and Supply Chain Assurance*, 2015.
[21] C. Cid, S. Murphy, and M. Robshaw, *Algebraic aspects of the advanced encryption standard*. 2006.
[22] Z. Wang, A. C. Bovik, H. R. Sheikh, and E. P. Simoncelli, "Image quality assessment: From error visibility to structural similarity," *IEEE Trans. Image Process.*, vol. 13, no. 4, pp. 600–612, 2004.
[23] Z. Wang and A. C. Bovik, "Mean Squared Error: Love It or Leave It?," *IEEE Signal Process. Mag.*, vol. 26, no. January, pp. 98–117, 2009.
[24] Z. Wang, A. C. Bovik, H. R. Sheikh, and E. P. Simoncelli, "The SSIM Index for Image Quality Assessment." [Online]. Available: https://ece.uwaterloo.ca/~z70wang/research/ssim/.

Topological Insulators:
Electronic Structure, Material Systems and Applications

Parijat Sengupta

Photonics Center, Boston University, Boston, MA, 02215, USA
Dept of Electrical and Computer Engineering, Purdue University,
West Lafayette, IN 47907, USA
parijats@bu.edu

Received 6 September 2016
Accepted 10 April 2017

Topological insulators are a new class of materials characterized by fully spin-polarized surface states, a linear dispersion, imperviousness to external non-magnetic perturbations, and a helical character arising out of the perpendicular spin-momentum locking. This article answers in a pedagogical way the distinction between a topological and normal insulator, the role of topology in band theory of solids, and the origin of these surface states. Numerical techniques including diagonalization of the TI Hamiltonians are described to quantitatively evaluate the behaviour of topological insulator states. The Hamiltonians based on continuum and tight binding approaches are contrasted. The application of TIs as components of a fast switching environment or channel material for transistors is examined through I-V curves. The potential pitfall of such devices is presented along with techniques that could potentially circumvent the problem. Additionally, it is demonstrated that a strong internal electric field can also induce topological insulator behaviour with wurtzite nitride quantum wells as representative materials.

1. Introduction

Traditionally, our understanding of different phases of matter such as solids and liquids relies on the associated symmetries of the system and the concept of order-parameter so widely-prevalent in modern condensed-matter physics. This idea is clearly seen, for instance, in the phase transition from the liquid state with rotational and translational symmetry into a crystal with discrete symmetries (e.g. translational, discrete rotational, inversion, etc.). More complex phases describable by the symmetry breaking paradigm include examples such as ferromagnets (rotational-symmetry breaking) and superconductors (broken-gauge symmetry). Despite the considerable success enjoyed by the Landau-Ginzburg theory of spontaneous-symmetry breaking, several notable exceptions exist that elude such a description. Historically, the integer quantum Hall effect (QHE) discovered in 1980 by Klaus Von Klitzing while examining the behaviour of electrons confined in two dimensions and subjected to a strong perpendicular magnetic field and an in-plane electric field is regarded as the first deviation from the symmetry-breaking theory. In this particular experiment, the measured Hall conductance turned out to be in exact quantized fundamental units of e^2/h.

$$\sigma = \nu e^2/h \qquad (1)$$

where ν assumes any integer value. Klaus von Klitzing also discovered that the two-dimensionally confined electron gas possesses a Hall resistance that shows a plateau like structure. This is remarkable not only because it deviates significantly from a classical linear Hall resistivity plot, the quantization described by Eq. (1) is independent of microscopic details, type of material, and purity of the experimental sample. Impurities and other imperfections do not change the Hall resistivity behaviour. The two-dimensionally confined electron gas when exposed to a strong magnetic field creates Landau levels with discrete energies forcing the electrons to move in a cyclotron orbit. The electrons executing harmonic oscillator motion are separated by energy levels $E_n = (n+1/2)\hbar\omega$ that allow the system to be an insulator when the Fermi energy is placed within the energy gap. Thus far, the quantum Hall sample is a true bulk insulator unless the chemical potential is aligned with one of the Landau levels. However, the experimental observation of the quantized nature of Hall conductivity is seemingly at odds with this picture. The quest to understand this remarkably precise Hall quantization has spawned theoretical developments that is at heart of the current world-wide focus on topological insulators. One of the primary accomplishments has been the recognition that the plateaus (which give the quantized conductance) have topological significance and cannot be explained by resorting to a description of the bulk electronic structure. It can be explained in terms of topological invariants, also known as Chern numbers.[1,2] The robustness of the integer quantum Hall phase arises because it is protected from being deformed into another phase with a different topology, identical in nature to way a torus is protected from being deformed into a sphere.[3] The Quantum Hall effect will not be elaborated in greater detail here; the interested reader can refer to extensive original literature available.[4-7]

The QHE edge states are now easily understood in terms of the Laughlin picture[8] and composite fermions[9] but it is the realization that such edge states can also be present in absence of a time reversal breaking magnetic field that has led to the new field of topological insulators. The key difference from quantum hall systems is the absence of a magnetic field and production of double-degenerate edge states -one for each spin- as opposed to single current carrying state in the Quantum Hall effect. In the following sections, a simple model will be presented which captures the essence of topological insulators in an otherwise insulating sample.

The first example of a topological insulator (more precisely, a Chern insulator) was suggested by Duncan Haldane[10,11] who proposed an ingenious way to demonstrate the integer quantum Hall effect in a honeycomb lattice of spinless electrons placed in a periodic magnetic flux. The total magnetic flux crossing the lattice is zero but presence of an overall non-zero magnetic flux per unit cell drives the electrons to form an edge channel. The noteworthy feature of this proposal lies in the understanding that in absence of a net magnetic field, integer quantum Hall effect can be demonstrated. The quantized Hall conductance is therefore an outcome of the electron band structure for the lattice rather than the Landau states formed in presence of a finite external magnetic field. A crucial observation, as will be shown below, is that role of periodic magnetic flux can be replaced by the intrinsic spin orbit coupling.

2. The Kane Mele quantum spin Hall insulator

Until the prediction and experimental discovery of topological insulators, the quantum Hall effect in presence of a large magnetic field was the only known realization of topological state in existence. When compared to the rich variety of traditional broken-symmetry states, beginning from the ubiquitous solid-liquid and gaseous phases to superconductivity, one is led to the obvious question: should there not be other topological states, without a contribution from the time-reversal breaking magnetic field remaining to be discovered? The first answer to this question was provided by the independent theoretical prediction of Kane and Mele. In a pair of remarkable papers[12,13] in 2005, Kane and Mele motivated by the earlier work of Haldane put forward a theoretical prediction for a new state of matter; the 2D time-reversal invariant topological insulator or quantum spin Hall insulator (QSH). This state displays robust quantized properties but does not require a time-reversal symmetry breaking magnetic field for its observation. The remarkable proposal of Kane and Mele is based on the spin-orbit interaction of graphene.

The quantum spin Hall and the integer quantum Hall state are characterized by their chiral nature[14,15] though the preservation of time reversal symmetry in a quantum spin Hall insulator significantly alters the overall behaviour. In a QSH, two edge states which are related as time reversal symmetry pairs have opposite spins and propagate in clockwise and counter clockwise direction. Roughly speaking, the QSH whose origin lies in a spin orbit coupling induced momentum-dependent magnetic field can be viewed as two copies of the quantum Hall state with opposite Hall conductances, opposite spin, and spin-orbit coupling dependent magnetic field. However, the proposal by Kane and Mele to create a quantum spin Hall insulator turned out to be practically unrealisable because of the low spin-orbit coupling of graphene. Before introducing the theoretical prediction of QSH states in a HgTe/CdTe quantum well by Bernevig, Hughes, and Zhang, it is worthwhile to consider a few mathematical arguments to give this discussion a quantitative footing which will be useful later.

2.1. The role of spin-orbit coupling Hamiltonian

The Lorentz force, on account of the magnetic field which drives the quantum hall state and is the immediate theoretical precursor to quantum spin Hall state augments the Hamiltonian with a $\vec{A} \cdot \vec{P}$ orbital component term. Using a symmetric gauge, the vector potential $\vec{A}$ can be written as:

$$\vec{A} = \vec{B}/2(y, -x, 0) \tag{2}$$

where $\vec{B}$ is the applied external magnetic field. This gives a Hamiltonian of the form

$$H \propto (xp_y - yp_x) \tag{3}$$

Therefore, Bernevig et al. argued that the goal is to look for another force in nature which produces a similar Hamiltonian. The obvious candidate is the spin-orbit coupling force. Its Hamiltonian is of the form

$$H_{spin-orbit} = \vec{E} \times \vec{P} \cdot \sigma \tag{4}$$

where σ is the Pauli spin matrix. Instead of an external $\vec{B}$ field, an external $\vec{E}$ field is used which preserves time reversal symmetry. If one considers a $\vec{E}$ field of the form E($\vec{x}+\vec{y}$), the corresponding Hamiltonian becomes:

$$H_{spin-orbit} = \vec{E}\sigma_z(xp_y - yp_x) \tag{5}$$

The form of $\vec{E}$ considered is assumed to be confined in a two dimensional plane (along with the particle momentum), therefore only the z-component of spin enters the Hamiltonian. It is easy to notice that a heavy atom with a large nucleus will create a stronger electric field and the consequent spin-orbit interaction would be enhanced. This enhanced spin orbit interaction can lead to significant changes in the overall dispersion in a crystal as shown next by considering two relatively well known compound semiconductors, GaAs and HgTe.

2.2. The significance of band inversion

The concept of band inversion is easy to demonstrate by examining the ordering of bands in a zinc blende semiconductor which has a T_d point group symmetry. The highest valence and lowest conduction band state is made up of p and s orbitals respectively. A normal band order at the Γ point has the lowest conduction band state($j = 1/2$) with Γ_6 symmetry above the highest valence band state($j = 3/2$) with Γ_8 symmetry. In a normal ordered material, Γ_6 state is energetically higher than the Γ_8 state. If this order is reversed in bulk, say at the Γ point where the low-energy valence state(Γ_8) is pushed upwards such that the lowest conduction state(Γ_6) is energetically lower, an inverted band structure is created. In other words, the order of the high-symmetry conduction and valence bands is flipped. The flipping is most possible in crystals with a high degree of spin-orbit coupling. The energy gap at Γ which is usually defined as

$$E_g = E(\Gamma_6) - E(\Gamma_8), \tag{6}$$

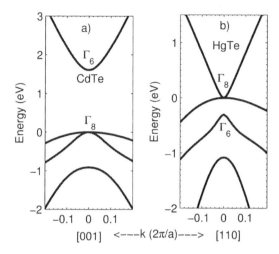

Fig. 1. Bulk band structure of CdTe(a) and HgTe(b). The ordering of the conduction and valence bands near the band gap at the Γ point in HgTe (Fig. 1b) is opposite to the one in CdTe (Fig. 1a). In HgTe, the hole state Γ_8 is above the electron state Γ_6 and band gap turns out to be negative.

The *normal* and *inverted* band structures of CdTe and HgTe are illustrated in Fig. 1(a) and Fig. 1(b) respectively.

In 2006, Bernevig, Hughes, and Zhang predicted[16] that HgTe quantum wells flanked by CdTe barriers would host quantum spin Hall insulator states along the edge. They based their prediction primarily on the inverted band structure of HgTe sandwiched between a normal ordered CdTe. The helical quantum spin Hall state was observed experimentally[17] in 2007 at the University of Wüzburg, Germany. At this point, it would be useful to introduce the idea of bulk boundary correspondence and how an inverted band structure is the starting point for the existence of a topological state.

3. Bulk boundary correspondence

This deep connection between the bulk and boundary which manifests itself in the form of robust surface or edge states can be intuitively understood in the following way. If we have a smooth interface between two materials with opposite signs for the bulk band-gap, the band structure changes slowly as a function of position across the interface. The energy gap therefore vanishes at the interface for a smooth transition from an inverted-order material with negative band gap to a normal-ordered positive band gap material. This brings us to the next pertinent question: Why are such states at the interface then so stable? The answer to this lies in time reversal symmetry, which is a fundamental symmetry present in nature. The interface state is always formed at what is called Time-Reversal-Invariant-Momenta(TRIM) points will be introduced.

In the bulk of a three dimensional material, time-reversal symmetry holds. This means that Eq. (7) is true.

$$E\left(\vec{k},\uparrow\right) = E\left(-\vec{k},\downarrow\right) \tag{7}$$

Additionally, if inversion symmetry holds the following relation is also true.

$$E\left(\vec{k},\uparrow\right) = E\left(-\vec{k},\uparrow\right) \tag{8}$$

If both equation (7) and equation (8) are simultaneously satisfied, bands are spin degenerate at the same $\vec{k}$ point.

$$E\left(\vec{k},\uparrow\right) = E\left(\vec{k},\downarrow\right) \tag{9}$$

Inversion symmetry is satisfied at special points in the Brillouin zone and in absence of any net external magnetic field, these points satisfy Eq. (9). The two spin-states of the conduction and valence bands are spin-orbit split but remain degenerate at any TRIM point. If the bulk is inverted at Γ which is a TRIM point, the zero band gap interface state formed by joining of the conduction and valence bands must also be at Γ point on the surface or edge. But since the conduction and valence band states are stable at any TRIM, the interface state formed by joining of the degenerate conduction and valence band state must also remain stable. In the presence of an external magnetic field which destroys time symmetry, the system ceases to have such robust interface states. On a surface which is two dimensional there are four such TRIM points: $(\pi,0), (\pi,\pi), (0,\pi), (0,0)$.

There is another important property of topological insulator: The surface bands intersect the fermi-level an odd number of times between two TRIM points. A trivial insulator has an even number of crossings. This has led to the classification now known as the Z_2 number. The two cases between a trivial and topological insulator can be distinguished by defining an index

$$N_k = m \mod 2 \tag{10}$$

where N_k is the number of Kramers pair of edge states that cross the Fermi energy. The expression for the index given in Eq. (10) simply means $m + 2p$, where p is any integer. If N_k is even, then $m = 0$, whereas $m = 1$ corresponds to N_k is odd. Since there are only two possible values of m, m is termed as being a Z_2 invariant. Z_2 is the group with two elements, namely 1 and 0 and hence is the simplest non-trivial group.

4. Generalization of quantum spin Hall state to 3D-topological insulators

The successful demonstration of the quantum spin Hall state in a HgTe/CdTe system focused attention on the problem of uncovering similar topological states protected by time reversal symmetry in three dimensions. The counterpart of the one-dimensional helical edge states in the quantum spin Hall insulator are the two-dimensional states with spin perpendicularly locked to momentum present on the surface. The surface is essentially an interface between the normal ordered vacuum and inverted order bulk. Several key papers[18–20] pointed to the possible existence of surface states in a three dimensional system. Bi_xSb_{1-x} was the first predicted 3D topological insulator which was subsequently observed by the Hasan group at Princeton through ARPES experiments.[21,22] The most common and simple example of 3D topological insulators are Bi_2Se_3, Bi_2Te_3, and Sb_2Te_3. These compounds have a single topological insulator (Dirac cone) on their surface. Bi_2Se_3 has been widely studied because of an energy-gap of 0.3 eV which is larger than the energy scale at room temperature. A complete description of the electronic structure of these compounds by taking Bi_2Se_3 as an example is available in literature.[23] It suffices to state here that the strong spin-orbit-coupling in Bi_2Se_3 leads to band inversion at the Γ point.

5. Hamiltonians for topological insulators

A reasonably accurate determination of the topological surface states begins by setting up a Hamiltonian that reflects the underlying electronic structure. In this section tight-binding and k.p Hamiltonians for the 2D (QSH) and 3D topological insulators would be presented. The parameter sets for these Hamiltonians are determined through a fitting process using data obtained from first-principles calculation.

5.1. Four-band k.p method for 3D topological insulators

The dispersion of Bi_2Te_3, Bi_2Se_3, and Sb_2Te_3 films are calculated using a 4-band k.p Hamiltonian. The 4-band Hamiltonian[18] is constructed (Eq. 11) in terms of the four lowest low-lying states $|P1_z^+ \uparrow\rangle$, $|P2_z^- \uparrow\rangle$, $|P1_z^+ \downarrow\rangle$, and $|P2_z^- \downarrow\rangle$. Additional warping effects[24] that involve the k^3 term are omitted in this low-energy effective Hamiltonian.

$$H(k) = \epsilon(k) + \begin{pmatrix} M(k) & A_1 k_z & 0 & A_2 k_- \\ A_1 k_z & -M(k) & A_2 k_- & 0 \\ 0 & A_2 k_+ & M(k) & -A_1 k_z \\ A_2 k_+ & 0 & -A_1 k_z & -M(k) \end{pmatrix} \quad (11)$$

where $\epsilon(k) = C + D_1 k_z^2 + D_2 k_\perp^2$, $M(k) = M_0 + B_1 k_z^2 + B_2 k_\perp^2$ and $k_\pm = k_x \pm i k_y$. For Bi_2Te_3 and Bi_2Se_3, the relevant parameters are summarized in Table I. The band structure of a 20.0 nm thick Bi_2Se_3 quantum well with a single Dirac cone around zero eV on the energy scale is shown in Fig. 2. A hallmark of the surface bands of a 3D-topological insulator is their intrinsic complete spin-polarization and locking of the spin perpendicular to the momentum. Within the framework of the 4-band k.p Hamiltonian, the expectation value of the three spin-polarized operators is computed. The operators for the three spin-polarizations are given by Eq. (12) and Eq. (13) in the Pauli representation σ_i $\{i = x, y, z\}$.

$$S_x = \begin{pmatrix} 0 & 0 & 1 & 0 \\ 0 & 0 & 0 & 1 \\ 1 & 0 & 0 & 0 \\ 0 & 1 & 0 & 0 \end{pmatrix}; \quad S_y = \begin{pmatrix} 0 & 0 & -i & 0 \\ 0 & 0 & 0 & -i \\ i & 0 & 0 & 0 \\ 0 & i & 0 & 0 \end{pmatrix} \quad (12)$$

$$S_z = \begin{pmatrix} 1 & 0 & 0 & 0 \\ 0 & 1 & 0 & 0 \\ 0 & 0 & -1 & 0 \\ 0 & 0 & 0 & -1 \end{pmatrix} \quad (13)$$

The above matrices are written in the basis set ordered as $|P1_z^+ \uparrow\rangle$, $|P2_z^- \uparrow\rangle$, $|P1_z^+ \downarrow\rangle$, and $|P2_z^- \downarrow\rangle$. The expectation value for each spin-polarization operator is calculated in the usual way in Eq. (14).

$$\langle S_i \rangle = \int \psi^* S_i \psi d\tau \quad (14)$$

where $\{i = x, y, z\}$.

5.2. Twenty-band $sp^3d^5s^*$ tight binding Hamiltonian for 3D TIs

A twenty-band atomistic tight-binding Hamiltonian[a] that calculates the dispersion over the complete Brillouin zone also captures the single Dirac cone topological surface state at the Γ point. The quintuple layer crystal structure[27] is imported to a twenty band tight binding model for 3D topological insulators. All parameters for these calculations were obtained from a orthogonal tight-binding model with $sp^3d^5s^*$ orbitals, nearest-neighbour interactions, and spin-orbit coupling.[28] Dispersion relationship obtained from the tight binding model was spin resolved along the quantized growth axis to identify spin polarization of the bands. A MATLAB script used to obtain the spin-polarization of bands is included in Appendix. The warped Fermi-surface contour which has a characteristic snow-flake structure as verified by experiments is also reproduced with the tight-binding Hamiltonian. The snow-flake feature is on account of higher order k^3 terms that are naturally included in the Hamiltonian to cover the full dispersion spectrum of the Brillouin zone.

5.3. Two-band Dirac Hamiltonian for 3D TIs

Dispersion relationships for surface bands with linearly dispersing states on surface of a topological insulator can be modeled using a two-dimensional Dirac Hamiltonian.

$$H_{surf.states} = \hbar v_f (\sigma_x k_y - \sigma_y k_x) \quad (15)$$

Here v_f denotes *Fermi*-velocity and σ_i where $i = x, y$ are the Pauli matrices. The two-dimensional Dirac Hamiltonian is in principle sufficient to probe the surface states and the dispersion relationship conforms well with experimentally observed

---

[a]All tight binding and transport calculations were performed with the NEMO5 software.[25,26]

Table I. 4-band k.p parameters[23] for Bi_2Te_3 and Bi_2Se_3.

Parameters	Bi_2Te_3	Bi_2Se_3
M (eV)	3.4	3.2
A_1 (eV Å)	5.6	5.3
A_2 (eV Å)	10.5	10.8
B_1 (eV Å^2)	24.6	22.5
B_2 (eV Å^2)	30.1	28.9
C (eV)	37.8	34.0
D_1 (eV Å^2)	0	0
D_2 (eV Å^2)	0	0

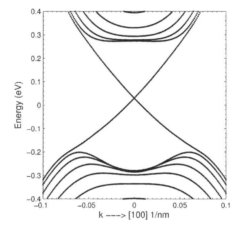

Fig. 2. Surface state formed in the mid-gap region of a 20.0 nm Bi_2Se_3 quantum well. The bulk band gap of Bi_2Se_3 is 0.32 eV at the Γ point. The conduction and valence bands are connected together by a cone of states at the Γ point. The meeting of the conduction and valence bands is usually a linear dispersion also known as a Dirac cone.

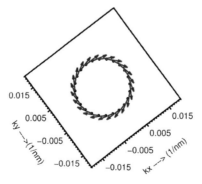

Fig. 3. The arrows show spin polarization confined to the plane in the vicinity of the Γ point for a 20.0 nm thick Bi_2Se_3 film. The spin is locked to momentum (which is a radial vector on the circle) shown by the tangential lines on the plot.

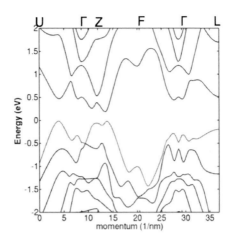

Fig. 4. Bulk band structure of Bi_2Te_3 according to parameterization of Ref. 28.

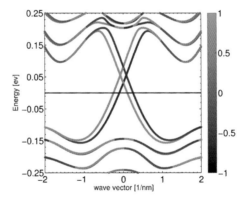

Fig. 5. The surface states of a topological insulator computed with the twenty-band tight binding method. The colour bar indicates the spin-polarization of the surface bands.

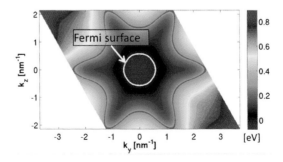

Fig. 6. The warping of the constant energy contour in to a snow-flake structure at energies away from the Dirac-point. This figure has been produced with a twenty-band tight binding model and matches well with the experimental data reported in literature.[29]

ARPES data. However, ARPES studies of the *Fermi*-surface of the topological insulator surface states at energies significantly above the Dirac point energy reveals a snow-flake like hexagram structure that is markedly different from a circular Fermi surface observed by application of Eq. (15). This departure from experiment was reconciled by noting that the simple two-dimensional Dirac Hamiltonian fails to account for the underlying crystal symmetries. The deformation of the Fermi surface can be established if higher order k terms are incorporated in the Hamiltonian. Since the two-dimensional Dirac Hamiltonian must comply with the C_{3v} point-group symmetry and time reversal symmetry, the next higher order terms that must be added are cubic in k. The modified Hamiltonian therefore must look like

$$H(k) = \epsilon_0(k) + \hbar v_f(\sigma_x k_y - \sigma_y k_x) + \frac{\lambda}{2}\left(k_+^3 + k_-^3\right)\sigma_z \quad (16)$$

$\epsilon(k)$ introduces the particle-hole anisotropy and the cubic terms denote warping. Using Eq. (16), the surface state spectrum becomes

$$\epsilon_\pm(k) = \epsilon_0(k) \pm \sqrt{\hbar v_f^2 k^2 + \lambda^2 \hbar^6 k^6 \cos^2(3\theta)} \quad (17)$$

The spectrum contains the lowest order correction to the perfect helicity of the Dirac cone predicted in Eq. (15). The $cos^2(3\theta)$ term possesses the six-fold symmetry due to C_{3v} symmetry and time reversal symmetry.

5.4. *Four-band k.p Hamiltonian for 2D TIs*

CdTe-HgTe-CdTe quantum wells which were the first predicted TIs are 2D topological insulators. Unlike their 3D TI counterpart, they possess bound states at the edge of the quantum well. The CdTe/HgTe/CdTe quantum well which behaves as a topological insulator beyond a critical well width dimension can be modeled using the BHZ Hamiltonian.[30] While an eight-band k.p Hamiltonian describes the full set of six valence (including spin split-off) and two conduction bands and their mutual interaction through the off-diagonal terms, it is sufficient to focus on bands that exclusively take part in the inversion process. This interaction of bands is governed by the coupling of conduction and valence states, represented through a linear term as shown in Eq. (18).

$$H(k) = \epsilon(k) + \begin{pmatrix} M_0 + M_2 k^2 & A(k_x + ik_y) & 0 & 0 \\ A(k_x - ik_y) & -M_0 + M_2 k^2 & 0 & 0 \\ 0 & 0 & M_0 + M_2 k^2 & A(-k_x + ik_y) \\ 0 & 0 & A(-k_x - ik_y) & -M_0 + M_2 k^2 \end{pmatrix} \quad (18)$$

where

$$\epsilon(k) = (C_0 + C_2 k^2) I_{4\times 4} \quad (19)$$

describes band bending. $2M_0 = -E_{g0}$ corresponds to energy gap between bands and is negative in the inverted order bands. This Hamiltonian is written in the basis of the lowest quantum well subbands $|E+\rangle$, $|H+\rangle$, $|E-\rangle$, and $|H-\rangle$. Here, $\pm$ stands for the two Kramers partners. The sign of the gap parameter M determines if it is a trivial insulator ($M > 0$) or a topological insulator ($M < 0$). Experimentally, M is tuned by changing the quantum well width. Experimentally, M_0 is tuned by changing the quantum well width. The parameters A, B, C, D, and M_0 are geometry dependent. For numerical calculations performed in this paper, the parameters were set to: $A = 364.5 meV.nm$, $B = -686 meV.nm^2$, $C = 0 meV$, $D = 512 meV.nm^2$, and $M = -10 meV$. A CdTe/HgTe/CdTe quantum well (Fig. 7) heterostructure with a well width under 6.3 nm exhibits a normal band order with positive E_g. Figure 7(a) confirms that the conduction states at Γ are indeed located above the valence states and the energy gap is positive. When the well width is exactly 6.3 nm, a Dirac system is formed in the volume of the device (Fig. 7). Beyond this *critical well width* of 6.3 nm, the heterostructure has its bands fully inverted. The band profile has a reverse ordering of the s-type and p-type orbitals (Fig. 8(c)) and $E_g < 0$. Accordingly, a nano-ribbon of width 100.0 nm formed by quantizing the quantum well in its in-plane direction has a positive band gap (as shown in Fig. 8(b)). Similarly, a nano-ribbon of width 100.0 nm constructed out of an inverted quantum well possesses gap-less TI edge states. The band structure of the nano-wire is illustrated in Fig. 8(d).

The corresponding absolute value of the squared edge-state wave functions is plotted in Fig. 9. The absolute value of the wave functions for the two edge states is maximum at the edge and gradually decay in to the bulk. This establishes that they belong exclusively to the edge states. The inversion of bands in the volume of the well is necessary for edge states with topological insulator behaviour. It is important to note however, that the process of inversion happens only at the Γ point. In the inverted dispersion plot (Fig. 8(c)), for momenta different from the Γ point, the band labeled with "H1" progresses from p to s-type. Similarly the band labeled with "E1" changes character from s to p. Both the bands, at a finite momentum acquire atomic orbital characteristics associated with a normally ordered set of bands. TI behavior is therefore restricted to a special set of TRIM points where the band structure is inverted.

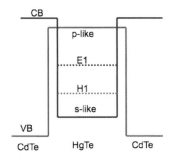

Fig. 7. Sketch of a CdTe/HgTe/CdTe quantum well heterostructure. The lowest conduction band (CB) state is labelled with E1 and the highest valence band (VB) state with H1.

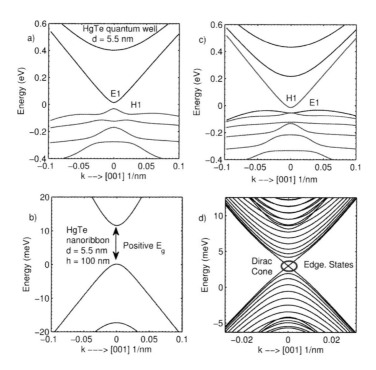

Fig. 8. Band structure of HgTe quantum well of thickness 6.3 nm. At this width, the lowest conduction band (E1) and highest valence band (H1) at the Γ point are equal.

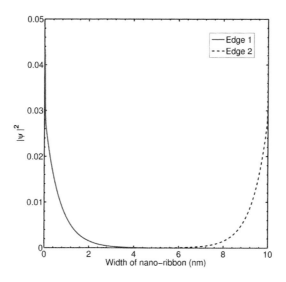

Fig. 9. Absolute value of the wave functions $|\psi|^2$ of the two edge-states of Fig. 8(d).

6. Topological insulator nanostructures

Topological insulator nanostructures offer significant advantages over bulk materials. The most crucial advantage of a nanostructure is the increased surface-to-volume ratio which is enhanced for a topological insulator because the surface states are the most important feature of these materials. Research groups worldwide have synthesised TI nanostructures.[31,32] This section presents information about TI-based devices that can be employed in the semiconductor industry. As a first example, the current-voltage characteristics of a TI ultra-thin body is examined. A possible use of this could be as components of a fast-switching low-power circuit environment. It is imperative to mention that TI surface states offer electronic mobility values that far outweigh those of Silicon, the current material extensively used in industry.[33,34] As a comparative measure, Si has mobility of 1400 cm^2/V.s while a Bi_2Te_3 thin film has been experimentally determined to have around 10,200 cm^2/V.s. The high mobility of TI surface states arise because of the protection it receives from back-scattering.

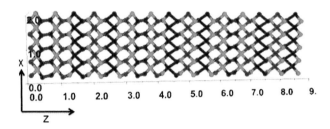

Fig. 10. Topological insulator ultra-thin body confined geometrically along z-direction and measures 8.942 nm. Contacts are placed along x-axis and this dimension is 1.972 nm. The y-axis is periodic.

An ultra-thin-body constructed out of Bi_2Te_3 is shown in Fig. 10. The structure is geometrically confined along the z-axis. Open boundary conditions (transport direction) is along the x-axis. The y-axis is assumed to be periodic. The temperature of operation is 300 K. An external voltage was applied between the two contacts placed along the x-axis. These two contacts serve as source and drain. The transmission profile of this device (Fig. 10) under the applied external voltage was computed within the non-equilibrium Green's function formalism.[35,36] Using the transmission data, current in the device as a function of the external voltage (Fig. 11) was computed by a direct application of the Landauer-Büttiker formalism. As a guide to the eye, the dispersion of the ultra-thin-body is shown in Fig. 12. External bias, with reference to Fig. 4 is applied between 0.0 and 0.2 eV to completely cover the TI surface states.

Similar current-voltage plots were compared with silicon ultra-thin body transistors. The current-voltage characteristics of silicon reveal that at low values of external voltages, current is negligible. A TI ultra-thin body on the other hand delivers sufficient current (Fig. 11) at a low source-drain bias. TI surface states can deliver current at low biases because of their semi-metallic character. The transistor fabricated out of a TI, therefore, is present in a turn-on state by default. Silicon, at low biases has zero or negligible density of states since they possess a finite band-gap

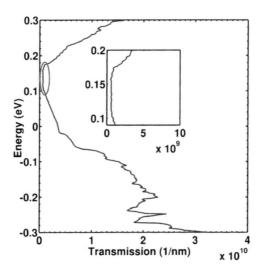

Fig. 11. Transmission profile of the TI ultra-thin body. The distinguishing feature of this transmission plot is the flat profile in the region of surface states.

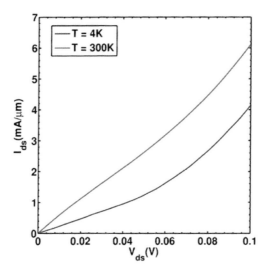

Fig. 12. I-V characteristics for the TI ultra-thin body shown in Fig. ??. At low bias values, the current delivered is sufficiently large compared to traditional semiconductor materials.

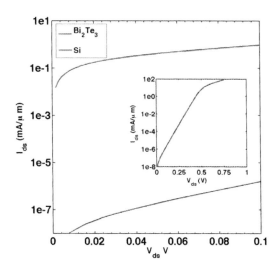

Fig. 13. Comparison of I-V characteristics of a Si and Bi_2Te_3 UTB. Current in Si UTB is negligible at low drain bias which is indicative of a natural transistor behaviour. The inset shows I_{on} current by Si UTB under a max. bias equal to 0.75V.

unlike the zero-gap metal-like states of TI. While, the aforementioned low-power attribute of TI based transistors is attractive, it also presents a significant challenge to turn-off this device. Leakage losses in a permanently turned-on device would far offset the potential benefits of a TI transistor. There are multiple ways, theoretically, to turn off a TI device. Under the action of an external magnetic field, the band-gap closing states would separate and form a finite gap. By placing the Fermi-level within the gap, the device can be turned off. In a real miniaturized semiconductor chip, an external magnetic field would be hard to apply without impacting the performance of the neighbouring electronic circuitry. A more sophisticated way of opening a band-gap is through a TI-superconductor heterostructure.

Before closing this section, it is noteworthy to point out that the surface termination of topological nanostructures play an important part in determining the overall electronic properties. A detailed account of such effects can be found elsewhere.[37] A symmetric thin film with identical surfaces, the two Dirac cones representing the surface states are degenerate. Asymmetry due to two different surfaces though can exist on account of inequivalent surface termination or presence of a substrate. Inversion symmetry therefore does not hold for such a TI film. An example of a real arrangement of atoms in a Bi_2Te_3 thin film with Bi and tellurium surface termination is shown in Fig. 14.[25] Under ideal conditions (ignoring impurity effects), a dipole is formed between the two surfaces as shown in Fig. 15. The corresponding charge fluctuation is also presented.

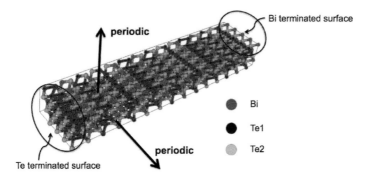

Fig. 14. Atomic structure of a Bi_2Te_3 thin film with two different surfaces. The two surfaces have Bi and Te termination thus making them chemically inequivalent.

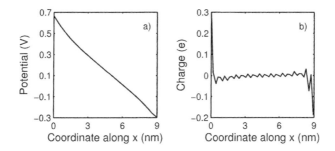

Fig. 15. The spatially-dependent electrostatic potential (Fig. 15a) and charge on each atomic node is plotted against the x coordinate of the Bi_2Te_3 thin film. This thin-film has Bi and Te termination on the surfaces. An oscillating charge pattern (Fig. 15b) is obtained for inequivalent surface termination.

7. Topological insulator-superconductor heterostructure

The previous section underscored the importance of a suitable mechanism to open a band gap and turn off a TI-based transistor. A simple approach would be to apply an external magnetic field or coat with a ferromagnet. The presence of a magnetic field will destroy time reversal symmetry and a band gap will develop; this approach though is fraught with serious shortcoming arising out of electromagnetic compatibility issues. A more sophisticated approach could be imparting superconducting properties to the topological insulator. It is now well established that doping of Cu or Pb can add superconducting states to Bi_2Se_3. The proximity effect at the interface between a superconductor and topological insulator has attracted considerable attention.[38,39] Because of proximity effect, when a TI surface state is closely placed with a superconductor, the superconductor's wave functions can penetrate into the topological surface states and turn them in to superconducting states. A superconductor has an intrinsic energy gap between the Fermi-level and the superconducting ground state. In this section, a modified version of the Bogoliubov-de Gennes Hamiltonian for a 3D-TI and s-wave superconductor is presented.

Before a complete Hamiltonian for a TI-superconductor heterostructure can be written, the conventional BCS description[40,41] of an s-wave superconductor must be examined. The BCS Hamiltonian in its simplest form can be written by starting with a Hamiltonian (Eq. 20) that describes a many-Fermion system interacting via a spin-independent interaction potential.

$$H = \sum_\sigma \int d^3 x \psi_\sigma^+(x) \left(\frac{-\hbar^2 \nabla^2}{2m} - \mu \right) \psi_\sigma(x)$$
$$+ \frac{1}{2} \sum_{\sigma\sigma'} \int d^3 x d^3 x' V(x,x') \psi_\sigma^+(x) \psi_\sigma^+(x') \psi_\sigma(x) \psi_\sigma(x') \tag{20}$$

In momentum space and finite volume, the following substitutions can be made

$$\psi(x) = \frac{1}{\sqrt{\Omega}} \sum_q e^{ik \cdot x}, V(x) = \frac{1}{\Omega} \sum_q e^{ik \cdot x} \overline{V_q} \tag{21}$$

The Hamiltonian in Eq. (20) can therefore be now written as

$$H = \sum_{k\sigma} \varepsilon_k a_{k\sigma}^+ a_{k\sigma} + \frac{1}{2\Omega} \sum_{\sigma\sigma'} \sum_{kk'q} \overline{V_q} a_{k+q,\sigma}^+ a_{k'-q,\sigma'}^+ a_{k\sigma} a_{k'\sigma'} \tag{22}$$

By restricting to paired fermions with zero total momentum and opposite spin, the BCS Hamiltonian can be written as

$$H = \sum_{k\sigma} \varepsilon_k a_{k\sigma}^+ a_{k\sigma} + \frac{1}{\Omega} \sum_{kk'} V_{k-k'} a_{k'\uparrow}^+ a_{-k'\downarrow}^+ a_{k\uparrow} a_{-k\downarrow} \tag{23}$$

The spectrum of this Hamiltonian when solved using the Bogoliubov transformation[42] yields a band structure with a gap in the spectrum.

For studying proximity effect between a superconductor and a topological insulator, the 4-band k.p model and the BCS Hamiltonian is used in conjunction. The fundamental assumption (experimentally verified) of the BCS Hamiltonian is the formation of Cooper pairs which are electrons with zero total momentum and spin. Superconductivity which is induced on the TI side of the TI-SC heterostructure must therefore agree to this principle. At this point it is worth mentioning again that the four orbitals participating in the electronic bonding process are $|P1_z^+ \uparrow\rangle$, $|P2_z^- \uparrow\rangle, |P1_z^+ \downarrow\rangle$, and $|P2_z^- \downarrow\rangle$. The composite Hamiltonian for the TI-SC structure similar to the BdG Hamiltonian can now be written as

$$H_{TS} = \begin{pmatrix} H_T - \mu & \Delta \\ \Delta^* & \mu - T H_T T^{-1} \end{pmatrix} \tag{24}$$

In Eq. (24), μ denotes the Fermi-level and T is the time reversal operator. H_{TS} is the composite Hamiltonian and H_T represents the 4-band k.p Hamiltonian. Δ is the pair potential given in the BCS formulation. For the s-wave superconductor considered here, the pair-potential is just a number. The analytic representation of pair-potential changes to a $\vec{k}$ dependent quantity if p or d-type superconductors are considered. The pair potential Δ connects the two electrons with opposite momentum and spin. For the case of a TI, which is turned in to a superconductor, the orbitals with opposite spin and momentum are paired. The two sets of orbitals in the 4-band TI Hamiltonian are therefore coupled by two pair potentials. The full TI-SC Hamiltonian H_{TS} in the basis set $|P1_z^+ \uparrow\rangle$, $|P2_z^- \uparrow\rangle$, $|P1_z^+ \downarrow\rangle$, $|P2_z^- \downarrow\rangle$, $-|P1_z^+ \uparrow\rangle$, $-|P2_z^- \uparrow\rangle$, $-|P1_z^+ \downarrow\rangle$, and $-|P2_z^- \downarrow\rangle$ can be now written as

$$H_{TS} = \begin{pmatrix} \epsilon+M & A_1 k_z & 0 & A_2 k_- & 0 & 0 & \Delta_1 & 0 \\ A_1 k_z & \epsilon-M & A_2 k_- & 0 & 0 & 0 & 0 & \Delta_2 \\ 0 & A_2 k_+ & \epsilon+M & -A_1 k_z & -\Delta_1 & 0 & 0 & 0 \\ A_2 k_+ & 0 & -A_1 k_z & \epsilon-M & 0 & -\Delta_2 & 0 & 0 \\ 0 & 0 & -\Delta_1^* & 0 & -\epsilon-M & A_1 k_z & 0 & A_2 k_- \\ 0 & 0 & 0 & -\Delta 2^* & A_1 k_z & -\epsilon+M & A_2 k_- & 0 \\ \Delta_1^* & 0 & 0 & 0 & 0 & A_2 k_+ & -\epsilon-M & -A_1 k_z \\ 0 & \Delta_2^* & 0 & 0 & A_2 k_+ & 0 & -A_1 k_z & -\epsilon+M \end{pmatrix} \quad (25)$$

where $\epsilon(k) = C + D_1 k_z^2 + D_2 k_\perp^2$, $M(k) = M_0 + B_1 k_z^2 + B_2 k_\perp^2$ and $k_\pm = k_x \pm i k_y$. In the above Hamiltonian, the Fermi-level μ has been set to zero. The dispersion for the TI obtained in the presence of a superconductor is shown in Fig. 13. A 50.0 nm Bi_2Se_3 film was layered with an s-wave superconductor. The superconducting properties are assumed to extend up to 25.0 nm as shown experimentally.[43-45] The remaining half of the slab is pure Bi_2Se_3 and possesses regular 3D TI properties. The s-wave superconductor is assumed to have the material properties of Aluminium[46] whose order parameter($\Delta_1 = 0.34$ meV at $T = 0$ K). The band dispersion of the surface states for the Bi_2Se_3 film coupled to the superconductor is shown in Fig. 16(a). Since the superconductor extends only until half of the structure, the second surface still shows a Dirac cone while the top surface has an open band gap. This is shown in the zoomed in Fig. 16(b).

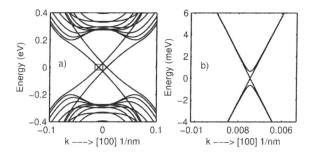

Fig. 16. The surface dispersion of a 50.0 nm Bi_2Se_3 film when coated with an s-wave superconductor. Figure 16(a) shows the overall band dispersion while Fig. 16(b) displays the energy dispersion around the Dirac cones. The box in Fig. 16(a) depicts an enlarged version of the Dirac cone split because of the superconducting proximity effect. The surface with no superconductor penetration has a TI surface state.

8. Summary and current trends

Topological insulators (TI) are a new class of materials whose surfaces host bound Dirac fermion like spin-polarized particles with high mobility. The discovery of topological insulators has opened up extensive experimental and theoretical research by several groups across the world. Potential future applications that include areas as diverse as thermoelectrics, spintronics, and quantum computing are being explored. It is beyond the scope of this review article to extensively discuss them as they on their own encompass a complete body of work. Comprehensive articles that deal in great detail several key aspects of topological insulators can be found in literature.[30,47,48]

Potential applications of topological insulators require synthesis of high-quality nanostructures such as thin-films and nanowires. The methods for synthesizing TI nanostructures include mechanical exfoliation from a bulk crystal and a vapour-solid or vapour-solid-liquid mechanism.[49] Despite the promise of a new class of devices based on topological surface states, the presence of bulk carriers that contaminate the surface pose the most serious challenge. The doped nanostructures also host multiple defects, which with the low band gap provide multiple channels for conduction thus reducing the scope of a surface-only based operation. In recent months, a particularly important advancement in fabrication of topological insulator nanostructures was made by Yong Chen and his group at Purdue University. They demonstrated[50,51] that a 3D topological insulator bismuth antimony tellurium selenide($BiSbTeSe_2$) conducts only at its surface and not at all in the bulk. This discovery could lead significant improvement in synthesis of topological insulator nanostructures.

The idea of pursuing topological quantum computation using models that rely on topological order was first proposed by Kitaev.[52] This premise inspired the search for the elusive Majorana fermions first proposed in 1937 in high-energy physics. A Majorana fermion is a fermion that is its own antiparticle. The zero-energy modes of Majorana fermions have been theoretically predicted and experimental evidence observed by the Delft group.[53] This exotic particle obeys non-Abelian statistics and can be manipulated for quantum computations that are remarkably immune to error. The Majorana zero modes have been reported at the interface of TIs with superconductors and can be used as a basis for a qubit for quantum computing. The topological protection afforded to the qubit imparts immunity to information loss from decoherence which currently stands as the foremost bottleneck in quantum computing progress.[54]

As of this writing, topological insulators are primarily an object of research to yield more information than what was thought possible from classical band theory. The power of topological invariants has been expanded to include topological superconductors, topological crystalline insulators, and Chern insulators. etc. New compounds are still being discovered that have unique topological signatures; nevertheless practical implementation of devices that exploit the key attributes of a topological insulators are still some distance away. The primary roadblock in this pursuit is the refinement and improvement of synthesis techniques that haven't kept pace with theoretical findings. A promising area of application could be spin devices; the intrinsic spin-momentum locking and spin polarized nature of the surface bands in conjunction with proximity to magnets and superconductors may open new avenues in this regard.

Acknowledgment

Computational resources from nanoHUB.org and support by National Science Foundation (NSF) (Grant Nos. EEC-0228390, OCI-0749140) are acknowledged. This work was also supported by the Semiconductor Research Corporation's (SRC) Nanoelectronics Research Initiative and National Institute of Standards & Technology through the Midwest Institute for Nanoelectronics Discovery (MIND), SRC Task 2141, and Intel Corporation. The work at Boston University was supported in part by the BU Photonics Center and U. S. Army Research Laboratory through the Collaborative Research Alliance (CRA) for MultiScale multidisciplinary Modeling of Electronic materials (MSME).

References

1. D. Sheng, Z. Weng, L. Sheng and F. Haldane, Physical Review Letters **97**, 036808 (2006).
2. A. M. Essin and J. Moore, Physical Review B **76**, 165307 (2007).
3. J. Stillwell, *Geometry of surfaces* (Springer, 1992).
4. A. H. MacDonald, Quantum Hall Effect: A Perspective. Series: Perspectives in Condensed Matter Physics, ISBN: 978-0-7923-0538-5. Springer Netherlands (Dordrecht), Edited by AH MacDonald, vol. 2 **2** (1989).
5. M. Stone, Annals of Physics **207**, 38 (1991).
6. T. Chakraborty and P. Pietiläinen, *The quantum Hall effects* (Springer, 1995).
7. S. M. Girvin, arXiv preprint cond-mat/9907002 (1999).
8. R. B. Laughlin, Physical Review B **23**, 5632 (1981).
9. J. K. Jain and R. Kamilla, *Composite fermions* (Cambridge University Press Cambridge, 2007).
10. F. D. M. Haldane, Physical Review Letters **61**, 2015 (1988).
11. B. A. Bernevig, *Topological Insulators and Topological Superconductors* (Princeton University Press, 2013).
12. C. L. Kane and E. J. Mele, Physical Review Letters **95**, 226801 (2005).
13. C. L. Kane and E. J. Mele, Physical Review Letters **95**, 146802 (2005).
14. L. Balents and M. P. Fisher, Physical Review Letters **76**, 2782 (1996).
15. C. Wu, B. A. Bernevig and S.-C. Zhang, Physical Review Letters **96**, 106401 (2006).
16. B. A. Bernevig, T. L. Hughes and S.-C. Zhang, Science **314**, 1757 (2006).
17. M. König, S. Wiedmann, C. Brüne, A. Roth, H. Buhmann, L. W. Molenkamp, X.-L. Qi and S.-C. Zhang, Science **318**, 766 (2007).
18. H. Zhang, C.-X. Liu, X.-L. Qi, X. Dai, Z. Fang and S.-C. Zhang, Nature Physics **5**, 438 (2009).
19. Y. Chen, J. Analytis, J.-H. Chu, Z. Liu, S.-K. Mo, X.-L. Qi, H. Zhang, D. Lu, X. Dai, Z. Fang et al., Science **325**, 178 (2009).

20. R. Roy, Physical Review B **79**, 195322 (2009).
21. Y. Xia, D. Qian, D. Hsieh, L. Wray, A. Pal, H. Lin, A. Bansil, D. Grauer, Y. Hor, R. Cava et al., Nature Physics **5**, 398 (2009).
22. D. Hsieh, Y. Xia, D. Qian, L. Wray, F. Meier, J. Dil, J. Osterwalder, L. Patthey, A. Fedorov, H. Lin et al., Physical Review Letters **103**, 146401 (2009).
23. C.-X. Liu, X.-L. Qi, H. Zhang, X. Dai, Z. Fang and S.-C. Zhang, Physical Review B **82**, 045122 (2010).
24. L. Fu, Physical Review Letters **103**, 266801 (2009).
25. J. E. Fonseca, T. Kubis, M. Povolotskyi, B. Novakovic, A. Ajoy, G. Hegde, H. Ilatikhameneh, Z. Jiang, P. Sengupta, Y. Tan et al., Journal of Computational Electronics **12**, 592 (2013).
26. https://engineering.purdue.edu/gekcogrp/software-projects/nemo5/.
27. W. Zhang, R. Yu, H.-J. Zhang, X. Dai and Z. Fang, New Journal of Physics **12**, 065013 (2010).
28. S. Lee and P. von Allmen, Applied physics letters **88**, 022107 (2006).
29. S. Basak, H. Lin, L. Wray, S.-Y. Xu, L. Fu, M. Hasan and A. Bansil, arXiv preprint arXiv:1103.4675 (2011).
30. X.-L. Qi and S.-C. Zhang, Reviews of Modern Physics **83**, 1057 (2011).
31. J. J. Cha, K. J. Koski and Y. Cui, physica status solidi (RRL)-Rapid Research Letters **7**, 15 (2013).
32. D. Kong, J. C. Randel, H. Peng, J. J. Cha, S. Meister, K. Lai, Y. Chen, Z.-X. Shen, H. C. Manoharan and Y. Cui, Nano letters **10**, 329 (2009).
33. D.-X. Qu, Y. Hor, J. Xiong, R. Cava and N. Ong, Science **329**, 821 (2010).
34. F. Xiu, L. He, Y. Wang, L. Cheng, L.-T. Chang, M. Lang, G. Huang, X. Kou, Y. Zhou, X. Jiang et al., Nature nanotechnology **6**, 216 (2011).
35. T. C. Kubis, Quantum transport in semiconductor nanostructures, PhD thesis, Universität München (2009).
36. H. Haug and A.-P. Jauho, *Quantum kinetics in transport and optics of semiconductors* (Springer, 2007).
37. P. Sengupta, T. Kubis, Y. Tan and G. Klimeck, arXiv preprint arXiv:1408.6274 (2014).
38. T. D. Stanescu, J. D. Sau, R. M. Lutchyn and S. D. Sarma, Physical Review B **81**, 241310 (2010).
39. I. Vobornik, U. Manju, J. Fujii, F. Borgatti, P. Torelli, D. Krizmancic, Y. S. Hor, R. J. Cava and G. Panaccione, Nano letters **11**, 4079 (2011).
40. M. Tinkham, *Introduction to superconductivity* (Courier Dover Publications, 2012).
41. P. De Gennes, Superconductivity of Metals and Alloys (Advanced Book Classics) (Addison-Wesley Publ. Company Inc, 1999).
42. A. L. Fetter and J. D. Walecka, Quantum theory of many-particle systems (Courier Dover Publications, 2003).
43. L. Fu and C. L. Kane, Physical Review Letters **100**, 096407 (2008).
44. Y. Hor, A. Williams, J. Checkelsky, P. Roushan, J. Seo, Q. Xu, H. Zandbergen, A. Yazdani, N. Ong and R. Cava, Physical Review Letters **104**, 057001 (2010).
45. L. A. Wray, S.-Y. Xu, Y. Xia, Y. San Hor, D. Qian, A. V. Fedorov, H. Lin, A. Bansil, R. J. Cava and M. Z. Hasan, Nature Physics **6**, 855 (2010).
46. K. Fossheim and A. Sudbø, *Superconductivity: physics and applications* (John Wiley & Sons, 2005).
47. Y. Ando, arXiv preprint arXiv:1304.5693 (2013).
48. M. Z. Hasan and C. L. Kane, Reviews of Modern Physics **82**, 3045 (2010).
49. X. Fa-Xian and Z. Tong-Tong, Chinese Physics B **22**, 96104 (2013).

50. Y. Xu, I. Miotkowski, C. Liu, J. Tian, H. Nam, N. Alidoust, J. Hu, C.-K. Shih, M. Z. Hasan and Y. P. Chen, Nature Physics (2014).
51. `http://physicsworld.com/cws/article/news/2014/nov/20/new-3d-topological-insulator-is-the-nearest-to-perfection-yet`.
52. J. Alicea, Y. Oreg, G. Refael, F. von Oppen and M. P. Fisher, Nature Physics **7**, 412 (2011).
53. V. Mourik, K. Zuo, S. Frolov, S. Plissard, E. Bakkers and L. Kouwenhoven, Science **336**, 1003 (2012).
54. J. K. Pachos, *Introduction to topological quantum computation* (Cambridge University Press, 2012).

Mosaic Crystal Model for Dynamical X-Ray Diffraction from Step-Graded In$_x$Ga$_{1-x}$As and In$_x$Al$_{1-x}$As/GaAs (001) Metamorphic Buffers and Device Structures

P. B. Rago*

Engineering Department, Phonon Corporation,
90 Wolcott Rd., Simsbury, CT 06070, USA
Electrical and Computer Engineering Department,
University of Connecticut,
371 Fairfield Way, Unit 2157, Storrs, CT 06269-2157, USA
*paul.rago@gmail.com

J. E. Ayers

Electrical and Computer Engineering Department,
University of Connecticut,
371 Fairfield Way, Unit 2157, Storrs, CT 06269-2157, USA
john.ayers@uconn.edu

In this paper we apply a mosaic crystal model for dynamical x-ray diffraction to step-graded metamorphic semiconductor device structures containing dislocations. This model represents an extension of the previously-reported phase-invariant model, which is broadly applicable and serves as the basis for the x-ray characterization of metamorphic structures, allowing determination of the depth profiles of strain, composition, and dislocation density. The new model has more general applicability and is more appropriate for step-graded metamorphic device structures, which are of particular interest for high electron mobility transistors and light emitting diodes. Here we present the computational details of the mosaic crystal model and demonstrate its application to step-graded In$_x$Ga$_{1-x}$As/GaAs (001) and In$_x$Al$_{1-x}$As/GaAs (001) metamorphic buffers and device structures.

Keywords: dynamical diffraction; x-ray diffraction; dislocations; mosaic crystal model.

1. Introduction

Recently we reported a phase-invariant model for dynamical x-ray diffraction from metamorphic semiconductor heterostructures containing dislocations[1]. That model is applicable to most semiconductor structures and serves as the basis for their characterization by high-resolution x-ray diffraction, allowing depth profiling of the strain, composition, and dislocation density. The phase-invariant model represents an extension of the traditional approaches to x-ray analysis, whereby dynamical[2-7] and kinematical[8-12] simulations have been used in conjunction with a curve-fitting procedure to extract the profiles of strain and composition, but are based on perfect, dislocation-free laminar

*Corresponding author.

crystals. This renders the analysis inapplicable to mismatched structures with dislocation densities greater than 10^6 cm^{-2}. Krivoglaz and Ryaboshapka[13], and Levine and Thomson[14] have analyzed the line profiles of Bragg peaks from crystals containing straight, parallel screw dislocations with precisely known atomic displacements. Though their treatment is mathematically rigorous, it is difficult to apply to real crystals for which the dislocations present irregular structures with a variety of Burgers vectors and orientations. For this reason the phase-invariant model was constructed by considering the ensemble of dislocations in a defected semiconductor heterostructure to be described by characteristic distributions associated with the ensemble average angular and strain broadening of the Bragg profiles by the dislocations. Use of the phase-invariant model is broadly applicable to metamorphic semiconductor heterostructures and extends the x-ray analysis to determination of the depth profile of the dislocation density as well as the strain and composition. However, the phase invariant model loses accuracy in certain structures containing either closely lattice matched layers or layers with very high threading dislocation densities[15]. To overcome these limitations, we have developed a mosaic crystal model for x-ray diffraction which is more appropriate for step-graded metamorphic semiconductor structures. In this paper we present the physical and computational details of the mosaic crystal model, and we demonstrate its application to step-graded In$_x$Ga$_{1-x}$As/GaAs (001) and In$_x$Al$_{1-x}$As/GaAs (001) metamorphic buffers and device structures.

2. Theory

The diffraction profile for a perfect, infinitely thick semiconductor crystal may be calculated by solution of the Takagi-Taupin equation[2-4] for dynamical diffraction,

$$-i\frac{dX}{dT} = X^2 - 2\eta X + 1, \tag{1}$$

in which X is the complex scattering amplitude and η is the deviation parameter. The resulting scattering amplitude is described by the Darwin-Prins formula[19]

$$X = \eta_S - Sign(\eta_S)\sqrt{\eta_S^2 - 1}, \tag{2}$$

where the deviation parameter for the substrate is given by

$$\eta_S = \frac{-(\gamma_0/\gamma_H)(\theta - \theta_{BS})\sin(2\theta_{BS}) - 0.5(1-\gamma_0/\gamma_H)\Gamma F_{0S}}{\sqrt{|\gamma_0/\gamma_H|} C\Gamma\sqrt{F_{HS}F_{\overline{H}S}}}, \tag{3}$$

Where θ_{BS} is the Bragg angle for the substrate and θ is the actual angle of incidence on the diffracting planes, F_{0S}, F_{HS} and $F_{\overline{H}S}$ are the substrate structure factors for the 000, hkl, and $\overline{h}\overline{k}\overline{l}$ reflections, respectively, C is the polarization factor, and $\Gamma = r_e\lambda^2/(\pi V)$, where r_e is the classical electron radius, 2.818×10^{-5} Å, λ is the x-ray wavelength, and V is the unit cell volume.

In an imperfect or nonuniform crystal, such as a semiconductor device heterostructure, we must consider the diffuse scattering, which is the scattering of x-rays associated with imperfections (point defects, line defects, planar defects) and nonuniformities in stoichiometry, composition, and strain. Over the last several decades, a number of workers have developed mathematically rigorous models for the diffuse scattering associated with line defects (dislocations) by assuming a regular geometry involving one or more straight dislocations in which all atomic displacements are known precisely (see for example Krivoglaz and Ryaboshapka[13] and Levine and Thomson[14]). That work, though detailed and rigorous, neglects the diffuse scattering arising from nonuniformities and cannot be readily applied to real crystals, which contain random networks of dislocations for which the atomic displacements are not known in detail.

Li et al.[16] developed a model for diffuse scattering in an epitaxial multilayer containing a small and uniform dislocation density which was given indirectly in terms of the average mosaic block size L and a mean disorientation angle δ. In their work, Li et al. calculated the dynamical diffraction profiles based on a perfect crystal multilayer and added to it an incoherent intensity associated with the dislocations. However, because their approach involved adding intensity to the perfect crystal profile, it is not able to predict the reduction in peak intensities for a sample with a high dislocation density. Moreover, their approach does not allow for a depth variation of the dislocation density, so it is not applicable to depth profiling of dislocations.

In other closely related research, Fewster[17] and Fewster, Holy, and Andrew[18] modeled reciprocal space maps for semiconductor multilayers using dynamical diffraction theory, assuming a uniform mosaic structure given in terms of L and δ. Their work accounts for the diffuse scattering in such a way that is not restricted to low dislocation densities, but they do not account for a depth variation of the dislocation density and their emphasis is reciprocal space maps rather than high-resolution x-ray diffraction profiles which we consider here.

Recently we reported a phase-invariant model for dynamical diffraction from multi-layered semiconductor heterostructures with dislocations using a laminar approach to allow for depth variations of composition and strain, similar to Li et al.[16] and Fewster et al.[17,18]. In this approach, the dislocation density is modeled directly, rather than through the mosaic block size and disorientation angle. A modified form of the Takagi-Taupin equation for a distorted crystal was used instead of using the additive approach of Li et al.[16]. Also, similar to previous work, the diffuse scattering associated with point and planar defects was neglected in the phase-invariant model[5-7]. In the following we will review the dislocation-free crystal model, the phase-invariant model for dislocated crystals, and finally the mosaic crystal model.

In the dislocation-free crystal model, a semiconductor heterostructure is divided into a number of sublayers or lamina for application of the Takagi-Taupin equation[2-4], and a recursion formula is applied to the N sublayers in the laminar structure. For the nth layer in the stack, the scattering amplitude at the top of the layer, X_n, is related to the scattering amplitude at the bottom of the layer, X_{n-1}, by

$$X_n = \eta_n + \sqrt{\eta_n^2 - 1}\frac{(S_{1n} + S_{2n})}{S_{1n} - S_{2n}}, \tag{4}$$

where

$$S_{1n} = \left(X_{n-1} - \eta_n + \sqrt{\eta_n^2 - 1}\right)\exp\left(-iT_n\sqrt{\eta_n^2 - 1}\right) \tag{5}$$

and

$$S_{2n} = \left(X_{n-1} - \eta_n + \sqrt{\eta_n^2 - 1}\right)\exp\left(iT_n\sqrt{\eta_n^2 - 1}\right), \tag{6}$$

η_n is the deviation parameter and the thickness parameter is

$$T_n = h_n \frac{\pi \Gamma \sqrt{F_{Hn} F_{\bar{H}n}}}{\lambda \sqrt{|\gamma_0 \gamma_H|}}, \tag{7}$$

where h_n is the thickness and F_{0n}, F_{Hn} and $F_{\bar{H}n}$ are the 000, hkl, and $\bar{h}\bar{k}\bar{l}$ structure factors for the nth sublayer. This model may be applied to semiconductor heterostructures which are pseudomorphic and therefore free of dislocations to determine their depth profiles of strain and composition by the iterative curve-fitting process described above.

An imperfect crystal is distorted by the presence of dislocations in two important respects, and the effect on the diffraction profile may be modeled by treating the crystal as a mosaic of uniform blocks. First, the mosaic blocks exhibit angular variations, which causes diffraction at different angles relative to the crystal surface. Second, localized strain about the dislocations gives rise to variations in interplanar spacings, which alter the Bragg angles for the blocks. These mosaic block variations each bear statistical distributions, which, for angular and interplanar spacing, are given by $P_\alpha(\theta)$ and $P_\varepsilon(\theta)$, respectively. Here we have assumed Gaussian distributions but the approach described here is generally applicable to other distributions as well.

In the phase-invariant model, these two effects are accounted for by modifying the deviation parameter to

$$\eta_n = \left\{\sqrt{|\gamma_0/\gamma_H|}C\Gamma\sqrt{F_{Hn}F_{\bar{H}n}}\int_{-\infty}^{\infty}\int_{-\infty}^{\infty}\frac{P_\alpha(\alpha)P_\varepsilon(\beta)d\alpha d\beta}{-(\gamma_0/\gamma_H)(\theta + \alpha - \theta_{Bn} - \beta)\sin(2\theta_{Bn}) - 0.5(1 - \gamma_0/\gamma_H)\Gamma F_{0n}}\right\}^{-1}, \tag{8}$$

where θ_{Bn} is the Bragg angle for the nth sublayer and θ is the actual angle of incidence on the diffracting planes, F_{0n}, F_{Hn} and $F_{\bar{H}n}$ are the structure factors for the 000, hkl, and $\bar{h}\bar{k}\bar{l}$ reflections, respectively, C is the polarization factor, and $\Gamma = r_e\lambda^2/(\pi V)$, where r_e is the classical electron radius, 2.818×10^{-5} Å, λ is the x-ray wavelength, and V is the unit cell volume. By a treatment based on the mosaic block model of a defected semiconductor layer, it has been shown[20,22] that the distribution $P_\alpha(\theta)$ may be modeled by a Gaussian distribution given by

$$P_\alpha(\theta) = (1/(\sigma_\alpha\sqrt{2\pi}))\exp(-\theta^2/(2\sigma_\alpha^2)), \tag{9}$$

where the standard deviation is

$$\sigma_\alpha = b\sqrt{\pi D}/2, \tag{10}$$

D is the dislocation density, and b is the Burgers vector for the dislocations. It has also been shown, on the basis of the mosaic block model, that for a symmetric reflection from a (001) semiconductor heterostructure with 60° dislocations the distribution $P_\varepsilon(\theta)$ may be modeled as a Gaussian distribution[21,22] given by

$$P_\varepsilon(\theta) = (1/(\sigma_\varepsilon\sqrt{2\pi}))\exp(-\theta^2/(2\sigma_\varepsilon^2)), \tag{11}$$

for which the standard deviation is

$$\sigma_\varepsilon = 0.127b\sqrt{D\left|\ln(2\times 10^{-7} cm\sqrt{D})\right|}\tan\theta_B. \tag{12}$$

Although the approximate phase-invariant model outlined above is applicable to many metamorphic semiconductor heterostructures, it does not account for phase differences between the crystallites of the mosaic. This leads to incorrect predictions with respect to certain dynamic effects including extinction and Pendellosung fringes. This renders the predictions of the phase-invariant model inaccurate for some technically important cases of metamorphic structures, and this was the motivation for developing a refined mosaic crystal model which is more generally applicable to metamorphic semiconductor device structures.

In the mosaic crystal model, the structure is considered to comprise a mosaic of $N_\alpha \times N_\beta$ crystallites in α and β space. The dynamic scattering amplitude is computed for each of these crystallites iteratively. For the nth lamina of one such crystallite, the deviation parameter is

$$\eta_{nij} = \frac{-(\gamma_0/\gamma_H)(\theta-\theta_{Bn}+\alpha_i-\beta_j)\sin(2\theta_{Bn})-0.5(1-\gamma_0/\gamma_H)\Gamma F_{0n}}{\sqrt{|\gamma_0/\gamma_H|}C\Gamma\sqrt{F_{Hn}F_{\overline{H}n}}}, \tag{13}$$

where $\alpha_i = -N_\sigma\sigma_\alpha + iN_\alpha\sigma_\alpha/2N_\sigma$ and $\beta_j = -N_\sigma\sigma_\beta + jN_\beta\sigma_\beta/2N_\sigma$, where N_σ is the number of standard deviations used in the two distributions and i and j are integers. The scattering amplitudes are calculated iteratively by

$$X_{nij} = \eta_{nij} + \sqrt{\eta_{nij}^2 - 1}\frac{(S_{1nij}+S_{2nij})}{S_{1nij}-S_{2nij}}, \tag{14}$$

where

$$S_{1nij} = \left(X_{nij-1}-\eta_{nij}+\sqrt{\eta_{nij}^2-1}\right)\exp\left(-iT_n\sqrt{\eta_{nij}^2-1}\right) \tag{15}$$

and

$$S_{2nij} = \left(X_{nij-1}-\eta_{nij}+\sqrt{\eta_{nij}^2-1}\right)\exp\left(iT_n\sqrt{\eta_{nij}^2-1}\right), \tag{16}$$

η_n is the deviation parameter and the thickness parameter is

$$T_n = h_n \frac{\pi \Gamma \sqrt{F_{Hn} F_{\bar{H}n}}}{\lambda \sqrt{|\gamma_0 \gamma_H|}}, \qquad (17)$$

where h_n is the thickness and F_{0n}, F_{Hn} and $F_{\bar{H}n}$ are the 000, hkl, and $\bar{h}\bar{k}\bar{l}$ structure factors for the nth sublayer. The x-ray diffraction profile is calculated by adding the intensity contributions from the $N_\alpha \times N_\beta$ crystallites in α and β space; therefore

$$I = \sum_i \sum_j |X_{Nij}|^2 \cdot W_{\alpha i} \cdot W_{\varepsilon j}, \qquad (18)$$

where the weighting functions are given by

$$W_{\alpha i} = \exp(-\alpha_i^2 / 2\sigma_\alpha^2) \qquad (19)$$

and

$$W_{\varepsilon i} = \exp(-\beta_i^2 / 2\sigma_\varepsilon^2). \qquad (20)$$

This approach accounts for the phase incoherency exhibited by crystallites within the mosaic and may be applied to semiconductor heterostructures with arbitrary profiles of composition and dislocation density.

Although we have considered Gaussian distributions in this study, the dynamical models presented above may be readily applied using other angle-scale distributions, such as those suggested by Ivanov et al.[23] for metal crystals. Other distributions could be determined experimentally or by detailed calculations for particular configurations of dislocations using the theoretical framework of Krivoglaz and Ryaboshapka[13]. A future extension of this work might involve the application of rigorous models to random dislocation networks so that the relationships given by Eqs. (9)-(12) may be further refined. With ever-increasing computational power this may soon become feasible, but although success in this endeavor would enable the refinement of the models described by Eqs. (9)-(12), the overall approach would remain unchanged.

3. Results and Discussion

We have applied the mosaic crystal models for calculation of the x-ray diffraction profiles for example step-graded $InGa_xAs_{1-x}/GaAs$ (001) heterostructures. The x-ray wavelength was assumed to correspond to Cu $k_{\alpha 1}$ radiation (λ = 0.1540594 nm). In this work, the individual layers of the step-graded structure were assumed to have uniform dislocation density. Layers with constant composition but nonuniform dislocation density may be broken up into a series of sublayers for the purpose of x-ray simulation. We used N_σ = 3 and found that incorporation of a larger number of standard deviations did not provide an appreciable increase in accuracy. For implementation of the mosaic crystal model, the minimum number of crystallites we used involved N_α = 21 and N_β = 21, resulting in a model involving 441 crystallites. The required number of crystallites depends on the maximum dislocation density in the structure, so in general good practice requires that the number of crystallites be increased until the diffraction profile is unchanged by any increase in the number. For structures with an assumed dislocation density of $D = 10^8$ cm^{-2}, $N_\alpha = N_\beta = 31$ was used. To account for typical instrumental effects, calculated diffraction profiles may be convolved with an instrumental broadening function whose width depends on the geometry as well as the instrument. Here, however, we present diffraction profiles which are absent of instrumental effects.

First we considered 1.0-μm step-graded layers of In$_x$Ga$_{1-x}$As/GaAs (001), with the composition graded from 0% to 50% indium in five steps. Figure 1 shows the 004 diffraction profiles for this structure with five different assumed values of the dislocation density, and the perfect case. Although five diffraction peaks are easily resolved in the dislocation-free structure, these diffraction peaks broaden and overlap increasingly as the dislocation density increases. In addition, Pendellosung (diffraction interference fringes) tend to wash out with $D > 10^7$ cm^{-2}.

Next we investigated step-graded structures with ten steps instead of five, but with the same thickness and final composition as the previous set. The 004 diffraction profiles for these structures are shown in Fig. 2, for five different uniform dislocation densities. Here it is seen that the ten diffraction peaks associated with the compositions in the step-graded structure are increasingly difficult to resolve at the higher dislocation densities.

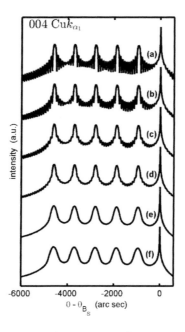

Fig. 1. 004 Diffraction Profile for Step Graded InGaAs/GaAs (five steps) with Dislocation Density of (a) zero, (b) 10^6 cm^{-2}, (c) 5×10^6 cm^{-2}, (d) 10^7 cm^{-2}, (e) 5×10^7 cm^{-2}, and (f) 10^8 cm^{-2}.

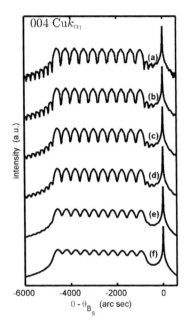

Fig. 2. 004 Diffraction Profile for Step Graded InGaAs/GaAs (ten steps) with Dislocation Density of (a) zero, (b) 10^6 cm^{-2}, (c) 5×10^6 cm^{-2}, (d) 10^7 cm^{-2}, (e) 5×10^7 cm^{-2}, and (f) 10^8 cm^{-2}.

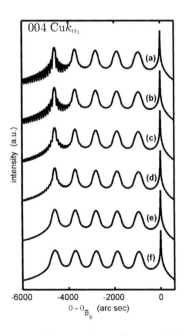

Fig. 3. 004 Diffraction Profile for Step Graded InGaAs/GaAs (five steps) with D = 10^8 cm^{-2} at the interface varied stepwise linearly to (a) zero, (b) 10^6 cm^{-2}, (c) 5×10^6 cm^{-2}, (d) 10^7 cm^{-2}, (e) 5×10^7 cm^{-2}, and (f) 10^8 cm^{-2} at the top layer. (Case (f) represents a uniform dislocation density.)

To study the effect of non-uniform dislocation density on the diffraction profile, we considered 1.0-µm step-graded layers of $In_xGa_{1-x}As/GaAs$ (001), with the composition graded from 0% to 50% indium in five steps, and with the dislocation density decreased stepwise, in a linear fashion, in the step-graded layers. The dislocation density in the first layer, deposited directly on the substrate, was assumed to be 10^8 cm^{-2} in each case, but the dislocation density in the top layer was varied as indicated in Fig. 3, with values of 10^8, 5×10^7, 10^7, 5×10^6, 10^6, and 0 cm^{-2}. Clear differences are seen in the diffraction profiles (with the greatest change in the diffraction peak corresponding to the top layer) indicating that the x-ray diffraction profile curve is sensitive to the profile of the dislocation density.

Much more work is necessary to comprehensively evaluate the application of the mosaic crystal model to experimental structures. As a preliminary study, we ran simulations comparing to a heterostructure grown by Jiang et al.[24] in a 2008 work. The structure comprises a GaAs substrate, a 900nm overshoot step-graded $In_xAl_{1-x}As$ buffer with nine steps from $x = 0.05$ to $x = 0.85$, and a 1 µm $In_{0.75}Al_{0.25}As$ top layer. Figure 4 shows the experimental 004 diffraction profile along with the diffraction profile calculated using the mosaic crystal model with the assumption that the structure is unstrained – as suggested by Jiang et al.[24] – and for uniform dislocation densities of 2×10^8 cm^{-2} and 10^9 cm^{-2} in the top layer. The buffer was assumed to have D = 10^9 cm^{-2} in both cases.

If the top 1 µm $In_{0.75}Al_{0.25}As$ layer is assumed to have D = 10^9 cm^{-2} then the predicted profile exhibits a width (~ 600") and a normalized intensity (normalized by dividing by the substrate peak intensity, ~ 0.06) which closely match those of the experimental profile as shown in Fig. 4(a). On the other hand, if the top $In_{0.75}Al_{0.25}As$ layer is assumed to have a larger (or smaller) dislocation density then the peak width will be overestimated (underestimated) while the normalized peak intensity will be underestimated (overestimated). As an example, Fig. 4(b) shows the case in which the top layer is assumed to have a dislocation density of D = 2×10^8 cm^{-2}. The resulting width and normalized intensity are 280" and 0.15, respectively. These observations suggest that the actual dislocation density in the top uniform layer is approximately 10^9 cm^{-2}.

Neither calculated diffraction profile achieves good agreement with the experimental profile associated with the graded buffer. First of all, the experimental profile exhibits shifts of the peak positions, and the most obvious explanation for this is the presence of residual strain. The peak shifts are also nonuniform, which might correspond to nonuniform strain. If this is the correct explanation, then the layers closest to the substrate appear to include tensile in-plane strain. However, the mismatch strain in the structure is compressive. Another possible explanation involves the tolerance in compositional control; that is, the layers closest to the substrate could be indium deficient compared to the target compositions. Yet another explanation is the presence of nonuniform crystallographic tilting in the individual layers. Strain, compositional, and tilting effects could be separated by the analysis of asymmetric as well as symmetric hkl reflections obtained at different azimuths. A second issue involves the disagreement in the peak intensities for the layers in the graded buffer. This may be explained partly by the noise floor in the experimental profile. It could relate partly to growth rate variations as well. The origin of the remaining

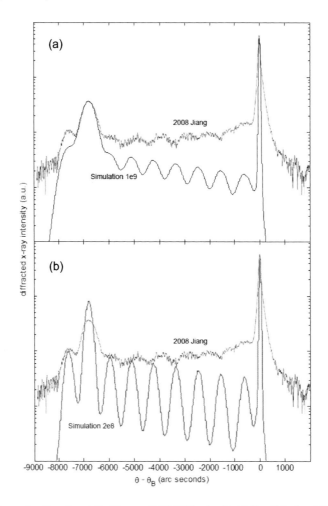

Fig. 4. Comparison of an Experimentally Measured 004 Diffraction Profile[24] to Calculated Diffraction Profiles with the dislocation density in the top layer assumed to be (a) $D = 10^9$ cm^{-2} and (b) $D = 2 \times 10^8$ cm^{-2}. The dislocation density in the graded buffer was assumed to be 10^9 cm^{-2} in both cases.

differences is unclear at this time. However, this work demonstrates the usefulness of the technique in estimating threading dislocation densities as well as compositions and strains in device structures. Additional information may be gleaned by further analysis using different hkl reflections and azimuths.

4. Conclusion

We have applied a mosaic crystal model for dynamical x-ray diffraction to step-graded $In_xGa_{1-x}As/GaAs$ (001) and $In_xAl_{1-x}As/GaAs$ (001) structures and compared calculated results to an experimental measurement. We show that the diffraction profile is sensitive to the dislocation density and its depth profile, so that in principle the x-ray diffraction profile analysis may allow depth profiling of the dislocation density as well as composition and strain in metamorphic semiconductor device structures if asymmetric and symmetric hkl high-resolution x-ray diffraction profiles are analyzed.

References

1. P. B. Rago, J. E. Raphael, and J. E. Ayers, *Materials Science and Technology Conference*, Montreal, Quebec (Oct. 27-31, 2013).
2. S. Takagi, *Acta Cryst.*, 15, 1311 (1962).
3. D. Taupin, *C. R. Acad. Sci.*, 256, 4881 (1963).
4. S. Takagi, *J. Phys. Soc. Jpn.*, 26, 1239 (1969).
5. M. A. G. Halliwell, M. H. Lyons, and M. J. Hill, *J. Cryst. Growth*, 68, 523 (1984).
6. C. R. Wie, T. A. Tombrello, and T. Vreeland, Jr., *J. Appl. Phys.*, 59, 3743 (1986).
7. W. J. Bartels, J. Hornstra, and D. J. W. Lobeek, *Acta Cryst.*, A42, 539 (1986).
8. V. S. Speriosu, *J. Appl. Phys.*, 52, 6094 (1981).
9. V. S. Speriosu and T. Vreeland, Jr., *J. Appl. Phys.*, 56, 1591 (1984).
10. L. Tapfer and K. Ploog, *Phys. Rev. B*, 40, 9802 (1989).
11. C. R. Wie, *J. Appl. Phys.*, 65, 1036 (1989).
12. C. R. Wie and H. M. Kim, *J. Appl. Phys.*, 69, 6406 (1991).
13. M. A. Krivoglaz and K. P. Ryaboshapka, *Fiz. Metal. Metalloved*, 15, 18 (1963).
14. L. E. Levine and R. Thomson, *Acta Cryst.*, 53, 590 (1997).
15. P. B. Rago and J. E. Ayers, *American Vacuum Society 60th International Symposium*, Long Beach, CA (Oct. 27 – Nov. 1, 2013).
16. M. Li, Z. Mai, J. Li, C. Li, and S. Cui, *Acta Cryst.*, A51, 350 (1995).
17. P. F. Fewster, *Conference on Optoelectronic and Microelectronic Materials and Devices*, pp. 1-8 (2000).
18. P. F. Fewster, V. Holy, and N. L. Andrew, *Mater. Sci. Semicond. Proc.*, 4, 475 (2001).
19. J. A. Prins, *Z. Phys.*, 63, 477 (1930).
20. P. Gay, P. B. Hirsch, and A. Kelly, *Acta Met.*, 1, 315 (1953).
21. M. J. Hordon and B. L. Averbach, *Acta Met.*, 9, 237 (1961).
22. J. E. Ayers, *J. Cryst. Growth*, 135, 71 (1994).
23. A. N. Ivanov, P. Klimanek, and A. M. Polyakov, *Metal Sci. Heat Tret.*, 42, 299 (2000).
24. Z. Jiang, W. Wang, H. Gao, L. Liu, H. Chen, J. Zhou, *Appl. Surf. Sci.*, 254, 5241 (2008).

Critical Layer Thickness:
Theory and Experiment in the ZnSe/GaAs (001) Material System

Tedi Kujofsa

Electrical and Computer Engineering Department,
371 Fairfield Way, Unit 4157,
Storrs, CT 06269-4157, USA
tedi.kujofsa@gmail.com

John E. Ayers

Electrical and Computer Engineering Department,
371 Fairfield Way, Unit 4157,
Storrs, CT 06269-4157, USA
john.ayers@uconn.edu

The critical layer thickness (CLT) determines the criteria for dislocation formation and the onset of lattice relaxation. Although several theoretical models have been developed for the critical layer thickness, experimentally-measured CLTs in ZnSe/GaAs (001) heterostructures are often at variance with one another as well as with established theories. In a previous work [T. Kujofsa *et al.*, J. Vac. Sci. Technol. B, 34, 051201 (2016)], we showed that the experimentally measured CLT may be much larger than the equilibrium value when using finite experimental resolution. In this work, we apply a general dislocation flow model to determine the apparent critical layer thickness as a function of the experimental resolution for ZnSe/GaAs (001) heterostructures. More importantly, we compare the results utilizing different equilibrium theories and therefore varying driving forces for the lattice relaxation in order to determine which established models are consistent with several measured values of CLT for ZnSe/GaAs (001) once kinetically-limited relaxation and finite experimental strain resolution are taken into account.

Keywords: apparent critical layer thickness; ZnSe/GaAs; experimental resolution.

1. Introduction

Understanding the critical layer thickness (CLT or h_c) for the introduction of misfit dislocations has important implications in the design and functionality of metamorphic devices. Although several models have been proposed for the critical layer thickness [1,2,3,4,5], the most commonly used and well-known model is that of Matthews and Blakeslee [1]. The critical layer thickness is the greatest thickness for which the equilibrium in-plane strain is equal to the lattice mismatch,

$$\varepsilon_{eq}(h) = f, \quad h \leq h_c.$$ (1)

In a previous work [6], we investigated the apparent critical layer thickness and the role of finite experimental resolution in the ZnSe/GaAs material system by considering the model of Matthews and Blakeslee (MB) for the equilibrium strain and the model of Kujofsa et al. [7] for the kinetically-limited strain. For epitaxially-grown ZnSe/GaAs (001), experimentally measured values of the critical layer thickness range from 50 nm to 225 nm [8,9,10,11,12,13,14,15] and these measurements show variations based on the growth method, temperature and characterization technique (i.e., experimental resolution). Although, in the ZnSe/GaAs (001) material system, the lattice mismatch and therefore the critical layer thickness vary with growth temperature owing to the difference in thermal expansion coefficients, such variation of the h_c from the often-quoted room-temperature value cannot explain the differences between the measured and equilibrium critical layer thickness values. Fritz [16] has argued that the differences between experimentally measured (apparent) critical layer thicknesses and equilibrium values stem from kinetically-limited relaxation combined with finite experimental resolution. In his work relating to SiGe and InGaAs, Fritz assumed that the lattice relaxation is a fixed fraction of the equilibrium relaxation ($\gamma / \gamma_{eq} = Q$), and showed that the measured critical layer thickness h_c^* may be much larger than the equilibrium value when using finite experimental resolution R, according to

$$f - R/Q = \varepsilon_{eq}(h_c^*).$$ (2)

Because kinetically-limited lattice relaxation is never a fixed fraction of the equilibrium strain, in our previous work [6], we considered the relationship between the resolution R and the apparent critical layer thickness h_c^* as

$$R = f - \varepsilon_{\|}(h_c^*).$$ (3)

Utilizing the kinetic-model described in Ref. 7, we determined the apparent critical layer thickness h_c^* at which the lattice relaxation can be detected using an experimental resolution R. A key result from that work was the finding that the Matthews and Blakeslee model is consistent with reported measurements of the critical layer thickness for ZnSe/GaAs (001). In this work, we use a similar approach to determine whether the experimental results are consistent with the models of van der Merwe (vdM) [2], People and Bean (PB) [3], Fischer et al. (Fis) [4] and Freund (Fre) [5]. For a more detailed description of this work, we refer the reader to Ref. 6. However, we will briefly summarize the procedure of this work below. We applied the kinetic model for lattice relaxation in order to predict the room-temperature residual strain in ZnSe/GaAs (001) as a function of thickness for given growth temperature conditions. Then, through use of Eq. (3), we calculated and plotted the apparent critical layer thickness as a function of experimental resolution for each of these growth temperature conditions (300 °C, 330 °C, 360 °C,

480 °C, 595 °C, and two-step growth where 40% of the thickness is grown at 595 °C and then the remaining 60% of the thickness is grown at 360 °C). Each measured value of the critical layer thickness was then plotted on the same graph according to the strain resolution of the particular measurement, using error bars to indicate the range of the experimental resolution in each case. The resolution range for each of the experimental points was determined by taking into account the effect of the counting statistics involved with determination of x-ray rocking curve peak position and width (for x-ray measurements). A more detailed description of this work could be found in Ref. 17. In such an analysis, if the experimental points fall upon the calculated curves, when accounting for the error bars, then we may conclude that the critical layer thickness model is consistent with the measurements.

2. Models for the Critical Layer Thickness and Equilibrium Strain

Although many theoretical models have been developed for three-dimensional deposits, or island growth, these will not be considered here and our main focus will be only on experimental results for continuous layers of ZnSe/GaAs (001), for which the Matthews and Blakeslee, van der Merwe, People and Bean, Fischer *et al.* and Freund models are applicable. The critical layer thickness expressions for all of the models considered here are summarized in Table 1.

Table 1. Summary of the critical layer thickness models.

Model	Critical Layer Thickness
Matthews and Blakeslee [1]	$h_c = \dfrac{b(1-v\cos^2\alpha)}{8\pi\|f\|(1+v)\cos\lambda}\left[\ln\left(\dfrac{h_c}{b}\right)+1\right]$
van der Merwe [2]	$h_c = \dfrac{a_e(1-2v)}{4\pi\|f\|(1-v)^2}\left[\ln(\|f\|)+\ln\left(\dfrac{2\pi}{e(1-v)}\right)\right]$
People and Bean [3]	$h_c = \dfrac{(1+v)b^2}{16\pi\sqrt{2}f^2(1-v)a_e}\ln\left(\dfrac{h_c}{b}\right)$
Fischer *et al.* [4]	$h_c = \dfrac{b\cos\lambda}{4\|f\|(1+v)}\left[1+\dfrac{(1-v\cos^2\alpha)}{4\pi(1+v)\cos^2\lambda}\ln\left(\dfrac{h_c}{b}\right)\right]$
Freund [5]	$h_c = \dfrac{b(1-v)\sin\alpha\tan\alpha}{8\pi\|f\|(1+v)\cos\lambda}\ln\left(\dfrac{h_c}{b}\right)$ $+ \dfrac{b\cos\alpha}{8\pi\|f\|(1+v)\cos\lambda}\left[\ln\left(\dfrac{h_c}{b}\right)-\dfrac{\cos 2\lambda}{2}-\dfrac{1-2v}{4(1-v)}\right]$

b is the length of the Burgers vector, v is the Poisson ratio, α is the angle between the Burgers vector and dislocation line vector, f is the lattice mismatch, λ is the angle between the Burgers vector and the direction in the interface which is perpendicular to the intersection of the glide plane and the interface and a_e is the lattice constant of the epilayer. In this work, we have assumed a room temperature (20 °C) lattice mismatch of ZnSe, $f = -0.270\%$, $\alpha, \lambda = 60°$, and $v = 0.38$.

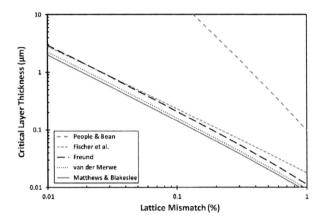

Fig. 1. Critical layer thickness as a function of the lattice mismatch for the models of Matthews and Blakeslee, van der Merwe, Freund, Fischer et al., and People and Bean.

Figure 1 compares the critical layer thickness as a function of the lattice mismatch for the various models considered here. Although all models show a monotonically decreasing CLT with increasing mismatch, there are wide departures amongst them. For the range of lattice mismatch investigated here, the Matthews and Blakeslee model yields the lowest critical layer thickness and therefore in the below analysis we will consider its results when differentiating with the other models; it should also be mentioned that there is no particular bias towards the Matthews and Blakeslee's model and its choice is made out of pure convenience. The van der Merwe characteristic is similar to the Matthews and Blakeslee curve and for the mismatch range considered in this work this model shows less than 15% difference in the prediction of the CLT compared to Matthews and Blakeslee. Although the Fischer et al. and Freund models exhibit similar expressions to the Matthews and Blakeslee model, they yield values which are higher by 25 to 50%. It should be mentioned that at low mismatch values (< 0.1%), the CLT characteristics for these two models overlap one another. However, at higher mismatch, the Freund model approaches the van der Merwe curve. The People and Bean critical layer thickness model was attractive in earlier work because it predicted fair agreement with experimental results for SiGe; however, in low mismatched material systems it greatly overestimates the critical layer thickness. In ZnSe/GaAs, for example, the People and Bean model predicts a critical layer thickness of ~15 μm which is orders of magnitude higher than experimentally-determined values.

In order to utilize the Kujofsa et al. plastic flow model for the determination of the strain-relaxation characteristics, the expressions of Table 1 must be rearranged in terms of the equilibrium in-plane strain. It should be noted that a key assumption here is that at the critical layer thickness, the equilibrium strain is equal to the lattice mismatch. Table 2 shows the equilibrium in-plane strain expressions for the various models. As a consequence of the results shown in Fig. 1, it is expected that the equilibrium in-plane strain characteristics will be different amongst these models which in turn will affect the kinetically-limited lattice relaxation.

Table 2. Summary of the equilibrium in-plane strain for various models.

Model	Critical Layer Thickness
Matthews and Blakeslee [1]	$\varepsilon_{eq}(h) = \dfrac{f}{|f|} \dfrac{b(1-\nu\cos^2\alpha)}{8\pi h(1+\nu)\cos\lambda}\left[\ln\left(\dfrac{h}{b}\right)+1\right]$
van der Merwe [2]	$\varepsilon_{eq}(h) = \dfrac{f}{|f|} \dfrac{a_e(1-2\nu)}{4\pi h(1-\nu)^2}\left[\ln\left(\left|\varepsilon_{eq}(h)\right|\right)+\ln\left(\dfrac{2\pi}{e(1-\nu)}\right)\right]$
People and Bean [3]	$\varepsilon_{eq}(h) = \dfrac{f}{|f|} \sqrt{\dfrac{(1+\nu)b^2}{16\pi\sqrt{2}h(1-\nu)a_e}\ln\left(\dfrac{h}{b}\right)}$
Fischer et al. [4]	$\varepsilon_{eq}(h) = \dfrac{f}{|f|} \dfrac{b\cos\lambda}{4h(1+\nu)}\left[1+\dfrac{(1-\nu\cos^2\alpha)}{4\pi(1+\nu)\cos^2\lambda}\ln\left(\dfrac{h}{b}\right)\right]$
Freund [5]	$\varepsilon_{eq}(h) = \dfrac{f}{|f|}\left(\begin{array}{l}\dfrac{b(1-\nu)\sin\alpha\tan\alpha}{8\pi h(1+\nu)\cos\lambda}\ln\left(\dfrac{h}{b}\right) \\ +\dfrac{b\cos\alpha}{8\pi h(1+\nu)\cos\lambda}\left[\ln\left(\dfrac{h}{b}\right)-\dfrac{\cos 2\lambda}{2}-\dfrac{1-2\nu}{4(1-\nu)}\right]\end{array}\right)$

The term $f/|f|$ accounts for the sign of the lattice mismatch and the equilibrium in-plane strain.

3. Results and Discussion

Figures 2(a), (c), (e), and (g) illustrate the calculated in-plane strain for the various models investigated here as a function of thickness for ZnSe/GaAs (001) grown under different temperature conditions: 300 °C, 330 °C, 360 °C, 480 °C, 595 °C, and two-step growth, for which 40% of the thickness is grown at 595 °C and then the remaining 60% of the thickness is grown at 360 °C. For the heterostructures considered here, the kinetically-limited in-plane strain characteristic exhibits a four-regime relaxation behavior (pseudomorphic, sluggish, rapid and saturation). The visibility of the four regimes is strongly controlled by the available thermal budget for relaxation and such phenomena have been shown in temperature-graded ZnSe/GaAs [18] and ZnS$_y$Se$_{1-y}$/GaAs [19] heterostructures. More importantly, ZnSe layers grown at higher temperatures exhibit a much more rapid decrease in the residual strain with thickness compared to those grown at lower temperatures. In addition, at relatively low temperature < 360 °C, due to the sluggish relaxation, a noticeable reduction in the residual strain does not become apparent until a thickness of ~500 nm. Because of this, measurement of the same apparent critical thickness will require much better experimental resolution with a growth temperature of 300 °C compared to the case of 595 °C. Not shown in Fig. 2 are the results from the consideration of the People and Bean model; because the critical layer thickness utilizing this particular model in ZnSe/GaAs (001) is ~15 μm, all the heterostructures considered here are below the critical thickness for dislocation formation and therefore according to the People and Bean model they are all predicted to be pseudomorphic layers which exhibit an in-plane strain of -0.27% at room temperature.

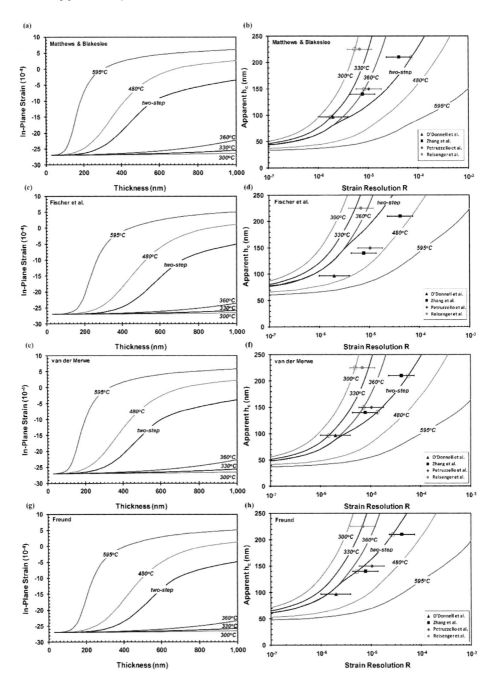

Fig. 2. (a, c, e, g) Calculated in-plane strain at room temperature as a function of thickness for ZnSe/GaAs (001) grown under different temperature conditions. (b, d, f, h) Apparent critical layer thickness for ZnSe/GaAs (001) as a function of the experimental resolution for different temperature conditions. The symbols represent experimental measurements and the horizontal error bars show the range of the estimated resolution. Each open circle indicates the intersection of the experimental point with the appropriate temperature curve.

Although the characteristic behavior of the residual strain is similar for the models of the Matthews and Blakeslee, van der Merwe, Fischer *et al.* and Freund, the numerical value of the strain differs at a given growth temperature and thickness. This variance in the residual strain stems from the difference in the equilibrium strain expressions given in Table 2. A key assumption of the plastic flow model is that the effective stress, which is the driving force of the lattice relaxation, is proportional to the difference of the actual and equilibrium strain. Therefore, a lower equilibrium strain will lead to a higher effective stress which in turn results in rapid relaxation rates and lower residual strain values. In comparison, for a given growth temperature and thickness the residual strain is lowest when considering the Mathews and Blakeslee model and highest utilizing the People and Bean model ($\varepsilon_{MB} < \varepsilon_{vdM} < \varepsilon_{Fre} < \varepsilon_{Fis} < \varepsilon_{PB}$).

From the results of Figs. 2(a), (c), (e), and (g), we created a plot of apparent critical layer thickness versus experimental resolution for each of the growth temperature conditions investigated, as follows. For each thickness and temperature condition, we found the extent of lattice relaxation observed at room temperature utilizing the kinetically-limited lattice relaxation model. An experimental method with a strain resolution equal to or less than (i.e., better than) this value would be able to detect the onset of lattice relaxation. Thus for the temperature of interest we may plot this thickness as the apparent critical layer thickness with a resolution equal to the expected lattice relaxation from Figs. 2(a), (c), (e), and (g). This yields a plot of apparent critical layer thickness as a function of experimental resolution, with growth temperature as a parameter, and this is given in Figs. 2(b), (d), (f), and (h) for all the various equilibrium models considered in this work. An important general finding from these plots is that, for the case of fixed resolution (for example, as determined by the counting statistics), an increase in the growth temperature is expected to lead to a lower determined h_c^*.

The experimental values for the apparent critical layer thickness considered here are 97 nm (O'Donnell *et al.* [11] for molecular beam epitaxy (MBE) at 330 °C using x-ray topography), 140 nm (Zhang *et al.* [8] for two-step metalorganic vapor phase epitaxy (MOVPE) growth using the x-ray full-width-at-half-maximum (FWHM) method), 150 nm (Petruzzello *et al.* [10] for 360 °C growth by MBE using the x-ray diffraction strain method), 210 nm (Zhang *et al.* [8] for two-step MOVPE growth using the x-ray strain method), and 225 nm (Reisinger *et al.* [9] for MBE growth at 300 °C using the x-ray diffraction strain method). For Zhang *et al.* [8], using the x-ray strain method, the strain resolution is estimated to be between 2.4×10^{-5} and 7.5×10^{-5}. For Zhang *et al.* [8] using the x-ray full-width-at-half-maximum method, the expected resolution is between 4.4×10^{-6} and 1.4×10^{-5}. For Petruzzello *et al.* [10], the expected resolution is between 5.9×10^{-6} and 1.9×10^{-5} for an effective number of counts between 2.0×10^{7} and 2.0×10^{9}. For Reisinger *et al.* [9], the expected resolution is between 3.9×10^{-6} and 1.2×10^{-5}. For O'Donnell *et al.* [11], the expected resolution is between 1.0×10^{-6} and 4.0×10^{-6}. For a complete description of the determination of the experimental resolution, we refer the reader to Ref. 6. The placement of the experimental point is determined by the geometrical mean $\bar{R}$ of the range of the experimental strain resolution:

$$\bar{R} = \sqrt{R_{min} R_{max}},\qquad(4)$$

where R_{min} and R_{max} are the limiting resolutions determined from the appropriate effective counts.

It is now possible to compare the five models for the critical layer thickness on the basis of Figs. 2(b), (d), (f), and (h). Figure 2(f) shows that the van der Merwe [2] model is consistent with all of the experimental data considered in this study and therefore appears to provide the most accurate description of the critical layer thickness and equilibrium strain, at least in the case of ZnSe/GaAs (001), from among the five models considered. The Matthews and Blakeslee [1] model is consistent with four out the five experimental data, but overestimates the critical layer thickness in the case of the Zhang et al. measurement based on the FHWM method. It should be noted at this point that we have not considered the uncertainty in film thickness due to growth rate variations or experimental error in the thickness characterization. If we had considered such thickness uncertainties, they would introduce vertical error bars. Then each data point would be plotted as a rectangle rather than a horizontal line segment. For the Zhang et al. data point based on the FWHM method, a thickness error of only 7% would cause this point to coincide with the Matthews and Blakeslee model. Considering this, we note that the Matthews and Blakeslee model provides a description which is nearly as accurate as the van der Merwe model, and may be applied to ZnSe/GaAs (001) for most practical purposes. On the other hand, the model of Freund [5] is consistent with only two of the experimental data points, and thickness tolerances of $\pm 15\%$ would have to be considered in order for this model to coincide with the experimental data. The model of Fischer et al. [4] is not consistent with any of the experimental data points, and thickness tolerances of $\pm 30\%$ would need to be introduced to account for the observed differences. Therefore we conclude that that these two models provide a less accurate description of the critical layer thickness and the equilibrium strain, compared to the van der Merwe and Matthews and Blakeslee models, in the case of ZnSe/GaAs (001).

4. Conclusion

Using a kinetic model for dislocation flow we have calculated the kinetically-limited lattice relaxation as a function of layer thickness for ZnSe/GaAs (001) layers grown under various temperature conditions. The effective stress, which is the driving force for lattice relaxation has been determined by considering various equilibrium models which include the Matthews and Blakeslee, van der Merwe, People and Bean, Freund and Fischer et al. On the basis of these lattice relaxation results, we have determined the apparent critical layer thickness as a function of the experimental resolution for each of these temperature conditions and CLT theories using a new model described by $R = f - \varepsilon_\parallel \left(h_c^* \right)$; that is, we have found the minimum thickness at which there will be detectable lattice relaxation using a given resolution. On the basis of this analysis, we conclude that the van der Merwe [2] and Matthews and Blakeslee [1] models are preferred for use in the ZnSe/GaAs (001) system. The models of Freund [5] and Fischer et al. [4] appear to provide less accurate descriptions of the critical layer thickness and equilibrium stain for ZnSe/GaAs (001). Finally, the People and Bean model greatly overestimates the critical layer thickness for this material system.

References

1. J. W. Matthews and A. E. Blakeslee, J. Cryst. Growth, 27, 118 (1974).
2. J. H. van der Merwe, J. Appl. Phys., 34, 123 (1962).
3. R. People and J. C. Bean, Appl. Phys. Lett., 49, 229 (1986).
4. A. Fischer, H. Kühne and H. Richter, Phys. Rev. Lett., 73, 2712 (1994).
5. L. B. Freund, J. Appl. Mech., 54, 553 (1987).
6. T. Kujofsa S. Cheruku, D. Sidoti, S. Xhurxhi, F. Obst, J. P. Correa, B. Bertoli, P. B. Rago, E. N. Suarez, F. C. Jain, and J. E. Ayers, J. Vac. Sci. Technol. B, 34, 051201 (2016).
7. T. Kujofsa, W. Yu, S. Cheruku, B. Outlaw, F. Obst, D. Sidoti, B. Bertoli, P. B. Rago, E. N. Suarez, F. C. Jain and J. E. Ayers, J. Electron. Mater. 41, 2993 (2012).
8. X. G. Zhang, P. Li, D. W. Parent, P. Li, G. Zhao, J. E. Ayers, and F. C. Jain, J. Electron. Mater., 28, 553 (1999).
9. T. Reisinger, M. J. Kastner, K. Wolf, E. Steinkirchner, W. Hackl, H. Stanzl, and W. Gebhardt, Mater. Sci. Forum, 182-184, 147 (1995).
10. J. Petruzzello, B. L. Greenberg, D. A. Cammack, and R. Dalby, J. Appl. Phys., 63, 2299 (1988).
11. C. B. O'Donnell, G. Lacey, G. Horsburgh, A. G. Cullis, C. R. Whitehouse, P. J. Parbrook, W. Meredith, I. Galbraith, P. Mock, K. A. Prior, and B. C. Cavanett, J. Cryst. Growth, 184/185, 95 (1988).
12. C. D. Lee, B. K. Kim, J. W. Kim, S. K. Chang, and S. H. Suh, J. Appl. Phys., 76, 928 (1994).
13. A. G. Kontos, E. Anastassakis, N. Chrysanthakopoulos, M. Calamiotou, and U. W. Pohl, J. Appl. Phys., 86, 412 (1999).
14. D. J. Olego, J. Vac. Sci. Technol. B, 6, 1193 (1988).
15. S. Ruvimov, E. D. Bourret, J. Washburn, and Z. Liliental-Weber, Appl. Phys. Lett., 68, 346 (1996).
16. I. J. Fritz, Appl. Phys. Lett., 51, 1080 (1987).
17. T. Kujofsa and J. E. Ayers, Int. J. Hi. Spe. Ele. Syst., (24), 1550007 (2015).
18. T. Kujofsa and J. E. Ayers, J. Electron. Mater. 44, 3030 (2015).
19. T. Kujofsa and J. E. Ayers, J. Electron. Mater. 45, 4580 (2016).

Author Index

Ahi, K. 1
Al Mamun, K. A. 25
Alhelal, D. 125
Althowibi, F. A. 55
Anwar, M. 1, 137
Ayers, J. E. 11, 55, 175, 187
Azadmehr, M. 69

Birge, R. R. 69

Chan, P.-Y. 99, 113
Chandy, J. 37, 99

Faezipour, M. 125
Fernandes, L. A. L. 69

Gogna, P. 37
Greco, J. A. 69

Häfliger, P. 69
Hasaneen, E-S. 37
Heller, E. 37, 99, 113
Hensley, D. K. 25

Islam, S. K. 25, 89

Jain, F. C. 37, 99, 113
Johannessen, E. A. 69

Kondo, J. 113
Kujofsa, T. 11, 187

Lingalugari, M. 99, 113

Mahbub, I. 89
Mazadi, A. 1
McFarlane, N. 25
Mirdha, P. 113

Rago, P. B. 175
Rivera, A. 1, 137

Saman, B. 37
Schwendemann, T. C. 49
Sengupta, P. 151
Shamsir, S. 89
Shanta, A. S. 25
Silliman, J. 49

Wagner, N. L. 69
Woods, K. 49